M338

The Open University

C1

Connectedness

This publication forms part of an Open University course. Details of this and other Open University courses can be obtained from the Student Registration and Enquiry Service, The Open University, PO Box 197, Milton Keynes, MK7 6BJ, United Kingdom: tel. +44 (0)870 333 4340, e-mail general-enquiries@open.ac.uk

Alternatively, you may visit the Open University website at http://www.open.ac.uk where you can learn more about the wide range of courses and packs offered at all levels by The Open University.

To purchase a selection of Open University course materials, visit the webshop at www.ouw.co.uk, or contact Open University Worldwide, Michael Young Building, Walton Hall, Milton Keynes, MK7 6AA, United Kingdom, for a brochure: tel. +44 (0)1908 858785, fax +44 (0)1908 858787, e-mail ouwenq@open.ac.uk

The Open University, Walton Hall, Milton Keynes, MK7 6AA.

First published 2006.

Copyright © 2006 The Open University

All rights reserved; no part of this publication may be reproduced, stored in a retrieval system, transmitted or utilised in any form or by any means, electronic, mechanical, photocopying, recording or otherwise, without written permission from the publisher or a licence from the Copyright Licensing Agency Ltd. Details of such licences (for reprographic reproduction) may be obtained from the Copyright Licensing Agency Ltd, 90 Tottenham Court Road, London W1T 4LP.

Open University course materials may also be made available in electronic formats for use by students of the University. All rights, including copyright and related rights and database rights, in electronic course materials and their contents are owned by or licensed to The Open University, or otherwise used by The Open University as permitted by applicable law.

In using electronic course materials and their contents you agree that your use will be solely for the purposes of following an Open University course of study or otherwise as licensed by The Open University or its assigns.

Except as permitted above you undertake not to copy, store in any medium (including electronic storage or use in a website), distribute, transmit or re-transmit, broadcast, modify or show in public such electronic materials in whole or in part without the prior written consent of The Open University or in accordance with the Copyright, Designs and Patents Act 1988.

Edited, designed and typeset by The Open University, using the Open University TeX System.

Printed and bound in the United Kingdom by The Charlesworth Group, Wakefield.

ISBN 0 7492 4134 9

Contents

Introduction	**4**
Study guide	4
1 Connected topological spaces	**5**
1.1 Disconnections and connectedness	5
1.2 Closures and disconnection	9
1.3 Unions of connected spaces	10
2 Connectedness, continuity and components	**11**
2.1 Connectedness and continuity	11
2.2 Components	13
3 Connectedness in Euclidean spaces	**16**
3.1 Connectedness in the real line	16
3.2 Homeomorphic subsets of the real line	19
3.3 Products of connected spaces	21
4 Path-connected spaces	**23**
4.1 What is a path?	23
4.2 Examples of path-connected spaces	26
4.3 Properties of path-connected spaces	30
5 The topologist's cosine	**32**
Solutions to problems	**37**
Index	**40**

Introduction

In Block A, we developed the general notions of metric and topological spaces, and considered several examples of such spaces. In Block B, we used these ideas to develop a theory and classification of surfaces. In doing so, we found it necessary to introduce extra conditions on the type of topological spaces that we allow: we required them to be path-connected, Hausdorff and compact. In this block, one of our aims is to explore more fully what these concepts entail.

See *Unit B1*, Section 1.

In this unit, we investigate what it means for a topological space to be *connected*. In doing so, we find that there are at least two distinct notions of topological connectedness. One is based on the notion of 'disconnectedness' — the idea that a topological space separates into at least two pieces, and then saying that any space that cannot be separated into 'distinct' pieces is connected (see Figure 0.1). As we would hope, the plane and the real line with the Euclidean topology are examples of spaces that are connected in this way. The other definition of connectedness is based on the idea of being able to find a 'path' between each pair of distinct points in the space.

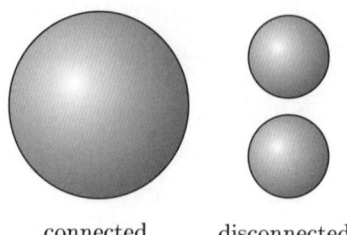

Figure 0.1

This is the notion of connectedness that we used in Block B.

Once we have these two notions of connectedness, we need to understand the relationship between them: if a space satisfies one definition of connectedness, does it automatically satisfy the other? It turns out that the notion of connectedness defined via paths always implies the notion defined via not being disconnected, but the converse is not always true. In Section 5, we give an example of a topological subspace of the Euclidean plane that is connected but not path-connected.

Path-connected $\Rightarrow$ connected; connected $\not\Rightarrow$ path-connected.

Study guide

The key ideas of this unit are contained in Sections 1, 2 and 4. These are the sections you should concentrate on if you are short of time. In Section 1 we define connectedness, in Section 2 we examine the relationship between connectedness and continuity, and in Section 4 we investigate path-connected spaces.

Section 3, *Connectedness in Euclidean spaces*, is an extended example, describing the connected subsets of $\mathbb{R}$. Of particular importance in this section is the generalization of the Intermediate Value Theorem.

Section 5, *The topologist's cosine*, contains an example which illustrates that path-connectedness does not imply connectedness, and that the closure of a path-connected set need not be path-connected. You should try to understand the ideas underlying the example, but you do not need to learn the technical details of the construction.

There is no software associated with this unit.

1 Connected topological spaces

After working through this section, you should be able to:
- define the terms *disconnection*, *disconnected* and *connected*;
- determine whether a given simple topological space is connected;
- produce disconnections of some disconnected spaces.

In this section we introduce our first definition of connectedness, by specifying what we mean by a *disconnected* topological space. We also look briefly at the relationship between closure and connectedness, and at unions of connected spaces.

1.1 Disconnections and connectedness

Our task is to answer the following question:

Given a topological space $(X, \mathcal{T})$, what should we mean by saying that X is disconnected?

We wish to end up with a notion that is topological (and so invariant under homeomorphisms of X), so we develop our ideas in terms of the open sets of X.

To inform our thinking, let us consider what happens when X consists of two disjoint intervals of the Euclidean line with the subspace topology. If the intervals are widely separated, then X should certainly be disconnected. If we move them towards each other, then eventually we come to a situation where we have to consider the topological properties of the intervals. If both are open at their neighbouring ends — for example $(0, 1)$ and $(1, 2]$ — then their union $X = (0, 1) \cup (1, 2]$ should be disconnected, because the point 1 is missing (Figure 1.1(a)). But if one contains the end point 1 — say the intervals are $(0, 1]$ and $(1, 2]$ — then their union X is the interval $(0, 2]$, which should be connected (Figure 1.1(b)).

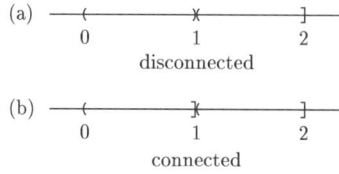

Figure 1.1

How can we use open sets to distinguish between these two cases? The key observation is that, in the first case, X is the union of two non-empty disjoint *open* sets (for the subspace topology on X inherited from the Euclidean topology on $\mathbb{R}$) — namely, the intervals $(0, 1)$ and $(1, 2]$. In the second case, when the intervals are $(0, 1]$ and $(1, 2]$, this is no longer the case: the intervals are also disjoint, but $X = (0, 2]$ and the interval $(0, 1]$ is not open (for the subspace topology on X). These observations underlie the following definition.

Definition

Let $(X, \mathcal{T})$ be a topological space. A **disconnection** $\{U, V\}$ of X is a pair of disjoint non-empty open subsets, U and V, with $X = U \cup V$.

The space $(X, \mathcal{T})$ is **disconnected** if X has a disconnection. It is **connected** if X has no disconnection.

A set $A \subseteq X$ is **connected** if $(A, \mathcal{T}_A)$ is connected. Otherwise it is **disconnected**.

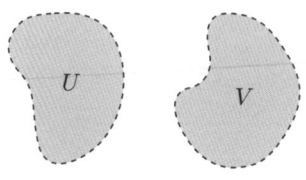

$X = U \cup V$ is disconnected

Figure 1.2

$\mathcal{T}_A$ denotes the subspace topology on A.

Remarks

(i) The intervals $(0, 1)$ and $(1, 2]$ form a disconnection of $(0, 1) \cup (1, 2]$ for the usual subspace topology, and so $(0, 1) \cup (1, 2]$ is disconnected.

(ii) It seems plausible that there are no disconnections of the interval $(0, 2]$ with the usual subspace topology. The proof of this involves results concerning the structure of $\mathbb{R}$ with the Euclidean topology, which we prove in Section 3.

(iii) When we wish to emphasize the topology, we say $\mathcal{T}$-*disconnection*, $\mathcal{T}$-*connected* and $\mathcal{T}$-*disconnected*.

(iv) Since U and V are required to be non-empty, neither U nor V can be X. Thus U and V must be proper non-empty subsets of X.

Recall that A is a *proper* subset of X if $A \neq X$.

Problem 1.1

Write down a disconnection of each of the following subsets of $\mathbb{R}$ when $\mathbb{R}$ has the Euclidean topology.

(a) $(-\infty, 0) \cup (1, \infty)$ (b) $(-\infty, 0] \cup (1, \infty)$ (c) $\{0\} \cup (1, 2]$

Problem 1.2

Show that the empty set, together with its (unique) topology, is a connected space.

Problem 1.3

Let $X = \{a, b, c\}$, and let $\mathcal{T}_1 = \{\varnothing, \{a\}, \{a, b\}, X\}$ and $\mathcal{T}_2 = \{\varnothing, \{a\}, \{b, c\}, X\}$ be topologies on X. Determine whether $(X, \mathcal{T}_1)$ and $(X, \mathcal{T}_2)$ are connected.

This example shows that whether a space is connected depends on the *topology*, as well as the set.

There is an important property of the sets U and V hidden in the definition of a disconnection. Because U is open, its complement $U^c = X - U = V$ is closed. So V is both closed and open. Similarly, U is both closed and open.

Lemma 1.1

Let $(X, \mathcal{T})$ be a topological space. If $\{U, V\}$ is a disconnection of X, then U and V are both clopen sets.

Recall, from *Unit A4*, that a *clopen* set is one that is both open and closed.

We saw in *Unit A4* that $\varnothing$ and X are always both open and closed subsets of X, and so every topological space has some clopen subsets. Lemma 1.1 tells us that a disconnected topological space always has at least two clopen subsets that are distinct from X and $\varnothing$.

We saw in Remark (iv) above that the sets forming a disconnection must be proper non-empty subsets of X.

It is natural to ask whether the converse also holds: if there are clopen subsets distinct from both $\varnothing$ and X, is the space disconnected? The answer is yes. This result is most easily phrased in terms of connectedness.

Theorem 1.2

The topological space $(X, \mathcal{T})$ is connected if and only if the only clopen subsets of X are $\varnothing$ and X.

Proof In Lemma 1.1, we saw that if X is disconnected then there are clopen sets other than $\varnothing$ and X. Hence if the only clopen subsets of X are $\varnothing$ and X, then $(X, \mathcal{T})$ is connected.

Suppose now that U is a non-empty and proper clopen subset of X: we must show that $(X, \mathcal{T})$ is disconnected. Define $V = U^c$. Since U is clopen, V is also clopen. Since U is proper, V is non-empty. Hence U and V are disjoint non-empty open sets and their union is X. Thus $\{U, V\}$ is a disconnection of X, and so $(X, \mathcal{T})$ is disconnected. ∎

Before continuing our discussion of the general theory of connected and disconnected spaces, we investigate whether some of our standard examples of topological spaces are connected or disconnected. Since producing a disconnection requires us to find two open sets that disconnect the set, we expect that the larger the topology, the more likely it is that a disconnection exists. Our first two examples bear out this intuition.

Worked problem 1.1 Let X be a set with at least two elements, and let $\mathcal{T}$ be the discrete topology on X. Show that $(X, \mathcal{T})$ is disconnected.

Solution

In order to show that $(X, \mathcal{T})$ is disconnected, we must find two disjoint non-empty open sets whose union is X. Since $\mathcal{T}$ is the discrete topology, all subsets of X are open. Let $x \in X$. Since X contains at least two points, $\{x\}^c \neq \varnothing$. Thus, the sets $\{x\}$ and $\{x\}^c$, being non-empty and open, form a disconnection of X. So $(X, \mathcal{T})$ is disconnected. ∎

Problem 1.4

Let X consist of a single element, and let $\mathcal{T}$ be the only possible topology that can be defined on X. Show that $(X, \mathcal{T})$ is connected.

Problem 1.5 Indiscrete topology

Let X be a set and let $\mathcal{T}$ be the indiscrete topology on X. Show that $(X, \mathcal{T})$ is connected.

Recall that the indiscrete topology on X consists of the two sets $\varnothing$ and X.

Consider a set X with at least two elements. With its indiscrete topology it is connected, and with its discrete topology it is disconnected. This is because in the indiscrete topology there are too few open sets available to form a disconnection, whereas in the discrete topology there are plenty. By using these two topologies as a guide, we come to the following result relating comparable topologies and connection.

Theorem 1.3

Let X be a set and let $\mathcal{T}_1$ and $\mathcal{T}_2$ be topologies on X with $\mathcal{T}_1 \subseteq \mathcal{T}_2$.
(a) If $(X, \mathcal{T}_1)$ is disconnected, so is $(X, \mathcal{T}_2)$.
(b) If $(X, \mathcal{T}_2)$ is connected, so is $(X, \mathcal{T}_1)$.

Proof The second statement is true if and only if the first statement is true, so we need to prove only the first.

Let $\{U, V\}$ be a $\mathcal{T}_1$-disconnection of X; since $\mathcal{T}_1 \subseteq \mathcal{T}_2$, U and V also belong to $\mathcal{T}_2$, and so $\{U, V\}$ is necessarily a $\mathcal{T}_2$-disconnection of X. ∎

Remark

This theorem does not cover all the possibilities that can arise:
- enlarging a topology *can* result in a change from connected to disconnected;
- diminishing a topology *can* result in a change from disconnected to connected.

The examples of the discrete and indiscrete topologies on a set X illustrate this (see Worked problem 1.1 and Problem 1.5).

However, the emphasis here is on the word *can* — these changes are possible, but not guaranteed: changing topologies may not change the connectedness.

Our next examples extend our understanding of connectedness in $\mathbb{R}$.

Worked problem 1.2

Determine whether $(\mathbb{Q}, \mathcal{T}_\mathbb{Q})$ is connected, when $\mathcal{T}_\mathbb{Q}$ is the subspace topology on $\mathbb{Q}$ inherited from $\mathbb{R}$ with the Euclidean topology.

Solution

Let x be any irrational number. Then $U = (-\infty, x) \cap \mathbb{Q}$ and $V = (x, \infty) \cap \mathbb{Q}$ form a $\mathcal{T}_\mathbb{Q}$-disconnection of $\mathbb{Q}$. So $(\mathbb{Q}, \mathcal{T}_\mathbb{Q})$ is disconnected. ∎

Note that $x \notin \mathbb{Q}$.

Problem 1.6

Determine whether $(\mathbb{R} - \mathbb{Q}, \mathcal{T}_{\mathbb{R}-\mathbb{Q}})$ is connected, when $\mathcal{T}_{\mathbb{R}-\mathbb{Q}}$ is the subspace topology on $\mathbb{R} - \mathbb{Q}$ inherited from $\mathbb{R}$ with the Euclidean topology.

Problem 1.7

Let $X = [-1, 1]$ carry the following topology:
$$\mathcal{T} = \{U \in X : 0 \notin U\} \cup \{U \subseteq X : (-1, 1) \subseteq U\}.$$
Show that $(X, \mathcal{T})$ is disconnected.

This example of an either-or topology was introduced in *Unit A3*, Problem 3.5.

Problem 1.8

Let X be a set with at least two elements and let $a \in X$. Show that X is connected for the a-deleted-point topology, $\mathcal{T}_a$.

Recall that, other than X, a set U is in $\mathcal{T}_a$ if and only if $a \notin U$.

So far, our only test that can prove directly that a space is connected is that $\varnothing$ and X are the only open sets — you used this in Problem 1.5. On the other hand, as you have seen, we can often discover a disconnection directly or, if the space is connected, prove that there cannot be one. Thus we usually work with disconnections.

1.2 Closures and disconnection

At the beginning of this section we considered the case when X consists of two intervals in $\mathbb{R}$. In particular, we considered the situation when these intervals approach each other. We now consider the analogous situation in the plane.

Consider two discs of radius 1 in the plane, situated so that they would meet at a single point were they both closed (see Figure 1.3). If both are open, their union is disconnected. If both are closed, then we expect their union to be connected. But what if one is open and the other is closed? Does this third case describe a connected or disconnected subspace of the plane? The answer is contained in the following lemma.

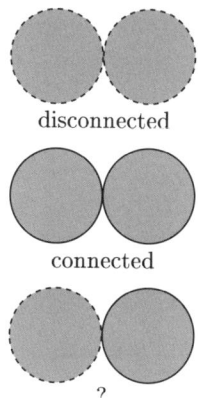

Figure 1.3

Lemma 1.4

Let $(X, \mathcal{T})$ be a topological space.

(a) If $\{U, V\}$ is a disconnection of X, then
$$\mathrm{Cl}(U) \cap V = \varnothing \text{ and } U \cap \mathrm{Cl}(V) = \varnothing.$$

(b) If there is a pair of non-empty sets U and V for which
$$U \cup V = X,\ \mathrm{Cl}(U) \cap V = \varnothing \text{ and } U \cap \mathrm{Cl}(V) = \varnothing,$$
then U and V are open and form a disconnection of X.

Recall from *Unit A4* (Theorem 2.5) that the closure $\mathrm{Cl}(A)$ of A is the smallest closed set that contains A.

Proof We prove (b), and leave the proof of (a) as a problem.

Suppose U and V are non-empty subsets of X, with
$$U \cup V = X,\ \mathrm{Cl}(U) \cap V = \varnothing \text{ and } U \cap \mathrm{Cl}(V) = \varnothing.$$

To show that U and V form a disconnection of X, we need to show that they are disjoint and open. Since $U \subseteq \mathrm{Cl}(U)$, we must have $U \cap V = \varnothing$, so U and V are disjoint. Suppose that U is not closed. Since its complement in X is V, there are points in its closure that belong to V. But that contradicts the hypothesis that $\mathrm{Cl}(U)$ is disjoint from V; hence U is closed. In a similar way, we deduce that V is closed.

Since U and V are complementary, the complement of U is V, and, since U is closed, V is open. Similarly, U is also open. Thus U and V form a disconnection of X. ∎

Problem 1.9

Prove Lemma 1.4(a).

We can now see that in the case when X consists of two just-touching discs in the plane with one open and the other closed, then (for the subspace topology) the discs do not give a disconnection of X, since the closed disc meets the closure of the open disc. (In fact, X is connected.)

1.3 Unions of connected spaces

We end this section by considering what happens when we take the union of a collection of connected spaces. In fact, the union of just two connected spaces can be disconnected: for example, in the real line with its usual topology, the intervals $[0,1]$ and $[2,3]$ are connected, but their union $[0,1] \cup [2,3]$ is not. In order to guarantee that the union of connected sets is connected, we require that the intersection of all the sets we are combining must be non-empty. This prevents any of the sets from being 'off on its own'.

> **Theorem 1.5**
>
> Let $(X, \mathcal{T})$ be a topological space, and let $\{A_i : i \in I\}$ be a family of connected subsets of X whose intersection $\bigcap_{i \in I} A_i$ is non-empty. Then $A = \bigcup_{i \in I} A_i$ is connected.

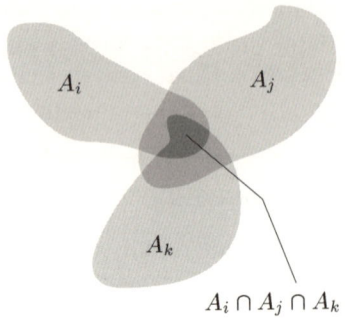

Figure 1.4

Proof Without loss of generality, we assume that I has at least two elements, since otherwise there is nothing to prove.

The proof is by contradiction. Assume that A is disconnected, and so there is a disconnection $\{U, V\}$ of A with respect to the subspace topology on A. Let a belong to the non-empty intersection of all the subsets A_i. Then $a \in A$ belongs to one of the sets of the disconnection — say $a \in U$.

Since V is non-empty, at least one of the sets A_i, the set A_j say, meets V, so the set $A_j \cap V$ is non-empty. Let $V_j = A_j \cap V$. Since $a \in A_j$ (a lies in every set of the family) and $a \in U$, $A_j \cap U$ is also non-empty. Let $U_j = A_j \cap U$.

Consider the pair U_j, V_j. We have just shown them to be non-empty. Since U and V are open in A, then U_j and V_j are open in A_j. They are also disjoint:

$$U_j \cap V_j = (A_j \cap U) \cap (A_j \cap V) = A_j \cap (U \cap V) = A_j \cap \varnothing = \varnothing.$$

Moreover, their union is A_j:

$$U_j \cup V_j = (A_j \cap U) \cup (A_j \cap V) = A_j \cap (U \cup V) = A_j \cap A = A_j.$$

So $\{U_j, V_j\}$ is a disconnection of A_j. But A_j is connected, so we have a contradiction. We conclude that A is connected. ∎

It is not true that the *intersection* of a family of connected sets must be connected. Indeed, the intersection of just two connected sets can be disconnected as the following problem illustrates.

Problem 1.10

Let

$$A = \{(x, y) \in \mathbb{R}^2 : x^2 + y^2 = 1, x \geqslant 0\}$$

$$B = \{(x, y) \in \mathbb{R}^2 : x^2 + y^2 = 1, x \leqslant 0\}.$$

(a) Sketch A and B, and find $A \cap B$.

(b) Show that, for the Euclidean topology on $\mathbb{R}^2$, $A \cap B$ is disconnected.

2 Connectedness, continuity and components

After working through this section, you should be able to:
▶ define connected spaces in terms of continuous functions;
▶ define the terms *component* and *totally disconnected*;
▶ describe the *components* of a given space.

In this section, we develop the theory of connected spaces by investigating how connectedness can be defined in terms of particular continuous maps. This approach allows simple proofs of various properties of connectedness — in particular:

- a continuous image of a connected space is connected;
- the largest connected set containing a given point is closed.

We also examine how the idea of the largest connected set that contains a given point can be used to split a set into disjoint connected components.

2.1 Connectedness and continuity

We now discuss the relationship between connectedness and continuous functions. Our first result gives a way of describing connectedness in terms of the existence of particular continuous functions.

We begin by considering the characteristic function which we introduced in *Unit A3*, Section 4. Recall that if $U \subseteq X$, then the characteristic function of U is the function $\chi_U \colon X \to \{0,1\}$ defined by

$$\chi_U(x) = \begin{cases} 1 & \text{if } x \in U, \\ 0 & \text{if } x \in U^c. \end{cases}$$

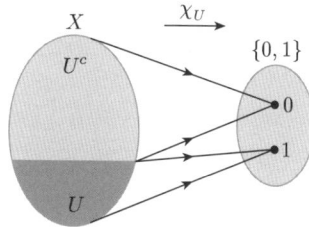

Figure 2.1

In *Unit A3*, we showed that, if $\mathcal{T}$ is a topology on X and $\mathcal{T}(d_0)$ is the discrete topology on $\{0,1\}$, then χ_U is $(\mathcal{T}, \mathcal{T}(d_0))$-continuous precisely when U and U^c both belong to $\mathcal{T}$.

Unit A3, Theorem 4.3.

More generally suppose that we can find a $(\mathcal{T}, \mathcal{T}(d_0))$-continuous function $f \colon X \to \{0,1\}$. Then

$$U = f^{-1}(\{1\}) \quad \text{and} \quad U^c = f^{-1}(\{0\})$$

are disjoint open sets. Thus X is disconnected, unless one of U and U^c is the empty set, in which case f is a constant function.

Here, f could be one of two constant functions: $f(x) = 0$ for all $x \in X$, or $f(x) = 1$ for all $x \in X$.

This gives the following useful characterization of connectedness.

Theorem 2.1

Let $(X, \mathcal{T})$ be a topological space. Then X is connected if and only if every $(\mathcal{T}, \mathcal{T}(d_0))$-continuous function $f \colon X \to \{0,1\}$ is constant.

Remark

One consequence of this theorem is that if X is connected, then no function from X to $\{0,1\}$ can be both continuous and *onto*.

By using this theorem we are able to prove the following important result which states that connectedness is preserved under *forward* mapping by a continuous function.

> **Theorem 2.2**
>
> Let $(X, \mathcal{T}_X)$ and $(Y, \mathcal{T}_Y)$ be topological spaces, let X be connected, and let $f \colon X \to Y$ be $(\mathcal{T}_X, \mathcal{T}_Y)$-continuous. Then $f(X)$ is connected.

Proof Let $\mathcal{T}_{f(X)}$ be the subspace topology that $f(X)$ inherits from $(Y, \mathcal{T}_Y)$. Notice that f is $(\mathcal{T}_X, \mathcal{T}_{f(X)})$-continuous.

This follows from *Unit A3*, Theorem 4.6.

Suppose now that $g \colon f(X) \to \{0,1\}$ is $(\mathcal{T}_{f(X)}, \mathcal{T}(d_0))$-continuous. Then the composite $g \circ f \colon X \to \{0,1\}$ is $(\mathcal{T}_X, \mathcal{T}(d_0))$-continuous. Hence, by Theorem 2.1, $g \circ f$ is constant. Now, if $a \in f(X)$, there exists $x \in X$ such that $f(x) = a$. Therefore $g(a) = g(f(x)) = (g \circ f)(x)$. Thus, since $g \circ f$ is constant, we deduce that g is constant.

This follows from *Unit A3*, Theorem 4.4.

Thus, since g is an arbitrary $(\mathcal{T}_{f(X)}, \mathcal{T}(d_0))$-continuous map from $f(X)$ to $\{0,1\}$, it follows that every $(\mathcal{T}_{f(X)}, \mathcal{T}(d_0))$-continuous map $g \colon f(X) \to \{0,1\}$ is constant. We conclude from Theorem 2.1 that $f(X)$ is connected. ∎

An important corollary of Theorem 2.2 is the fact that connectedness is a *topological invariant*: that is, if $(X, \mathcal{T})$ is connected, then any topological space *homeomorphic* to $(X, \mathcal{T})$ is also connected.

The concept of topological invariance was introduced in *Unit A4*.

> **Corollary 2.3**
>
> Connectedness is a topological invariant.

Proof Suppose that f is a homeomorphism from $(X, \mathcal{T}_X)$ to $(Y, \mathcal{T}_Y)$. Then f and f^{-1} are continuous functions, with $f(X) = Y$ and $f^{-1}(Y) = X$. If $(X, \mathcal{T}_X)$ is connected, then so is $(Y, \mathcal{T}_Y)$, by Theorem 2.2. Applying the same reasoning to f^{-1}, if $(Y, \mathcal{T}_Y)$ is connected, then so is $(X, \mathcal{T}_X)$. ∎

The converse of Theorem 2.2 is not necessarily true — we ask you to show this in the next problem.

Problem 2.1

Give an example of two topological spaces $(X, \mathcal{T}_X)$ and $(Y, \mathcal{T}_Y)$ and an onto $(\mathcal{T}_X, \mathcal{T}_Y)$-continuous function $f \colon X \to Y$ such that X is disconnected but Y is connected.

Another important property of connectedness is that it is preserved by the closure operation: if A is connected, then so is $\mathrm{Cl}(A)$. In fact, every subset of $\mathrm{Cl}(A)$ containing A is connected.

> **Theorem 2.4**
>
> Let $(X, \mathcal{T})$ be a topological space, let A be a connected subset of X and suppose that
>
> $$A \subseteq B \subseteq \mathrm{Cl}(A).$$
>
> Then B is connected.
>
> In particular, $\mathrm{Cl}(A)$ is connected.

Proof The proof is by contradiction. Assume that B is disconnected. Then Theorem 2.1 implies that there is a $(\mathcal{T}_B, \mathcal{T}(d_0))$-continuous function f from B onto $\{0, 1\}$. Hence $f|_A: A \to \{0, 1\}$, the restriction of f to A, is $(\mathcal{T}_A, \mathcal{T}(d_0))$-continuous. We obtain the desired contradiction by showing that $f|_A$ is onto.

This follows from *Unit A3*, Theorem 4.5.

Since f is onto, there is a point $b \in B$ such that $f(b) = 0$. Let $V = f^{-1}(\{0\})$. Then $V \in \mathcal{T}_B$, and so, by the definition of the subspace topology, there is a set $U \in \mathcal{T}$ such that $V = U \cap B$. Now U is a neighbourhood of b. Since $b \in B \subseteq \mathrm{Cl}(A)$, b is a closure point of A, and so every neighbourhood of b in X meets A. In particular, U has non-empty intersection with A, and so we can find $a \in U \cap A \subseteq V$. Then $a \in V$, and so $f(a) = 0$.

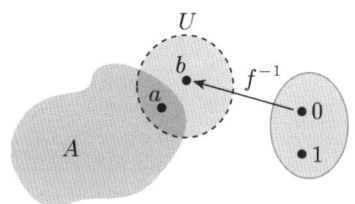

Figure 2.2

Since f is onto, there is also a point $b' \in B$ for which $f(b') = 1$. Employing the same reasoning as above, we conclude that there is a point $a' \in A$ for which $f(a') = 1$.

Thus f, when restricted to A, is also onto. Since $(A, \mathcal{T}_A)$ is connected, this contradicts the fact that no such continuous and onto function exists. Thus B is connected. ∎

Theorem 2.1.

This is a useful result, as it supplies us with many examples of connected sets. For example, once we have shown that the ball $B(0, 1)$ in the Euclidean plane is connected, we know that any set lying between $B(0, 1)$ and $B[0, 1]$ is also connected.

2.2 Components

Let $(X, \mathcal{T})$ be a topological space. A given point x in X is always a member of at least one connected subset of X — namely $\{x\}$. In fact, there is always a *largest* connected subset of X containing x, by which we mean a connected subset that contains all connected subsets containing x.

We saw that every one-element set is connected in Problem 1.4.

Definition

Let $(X, \mathcal{T})$ be a topological space, and let $x \in X$.

The **component** C_x of x in X is the largest connected subset of X that contains x.

X is **totally disconnected** if, for each point $x \in X$, the component of x is the set $\{x\}$.

The existence of C_x is proved below.

Remark

When we wish to emphasize the topology, we refer to the $\mathcal{T}$-component and say that X is totally $\mathcal{T}$-disconnected.

In Section 3, we show that, with respect to the Euclidean topology, every interval in $\mathbb{R}$ is connected. So, for example, if $X = [0, 1] \cup [2, 3]$ then $C_x = [0, 1]$ for each $x \in [0, 1]$, and $C_x = [2, 3]$ for each $x \in [2, 3]$.

We now show that the component of a point always exists and is unique.

> **Theorem 2.5**
>
> Let $(X, \mathcal{T})$ be a topological space, and let $x \in X$. Let
>
> $\mathcal{F}_x = \{A : x \in A, A \subseteq X, A \text{ is connected}\}.$
>
> Then C_x, the component of x, exists and is given by
>
> $C_x = \bigcup_{A \in \mathcal{F}_x} A.$

Proof As noted above, $\{x\} \in \mathcal{F}_x$ and so C_x is not empty. Moreover, the intersection of all the sets in $\mathcal{F}_x$ is non-empty, since each set contains the point x. Thus, by Theorem 1.5, $\bigcup_{A \in \mathcal{F}_x} A$ is connected.

If A is a connected subset of X that contains x, then $A \in \mathcal{F}_x$, and so $A \subseteq \bigcup_{A \in \mathcal{F}_x} A$. So $\bigcup_{A \in \mathcal{F}_x} A$ is the unique largest connected set that contains x — that is, $\bigcup_{A \in \mathcal{F}_x} A = C_x$. ∎

Note that, if X is connected, then $C_x = X$ no matter what $x \in X$ we choose. This tells us immediately that the component of any point in the following connected spaces is the whole space:

- any set carrying its indiscrete topology;   *See Problem 1.5.*
- any set containing at least two elements and carrying a deleted-point topology.   *See Problem 1.8.*

Problem 2.2

Describe the components of $X = \{a, b, c\}$ with respect to the topology $\{\varnothing, \{a\}, \{b, c\}, X\}$.

We now look at some examples of spaces that are totally disconnected.

Worked problem 2.1

Let $\mathbb{Q}$ have the subspace topology that it inherits from $\mathbb{R}$ with the Euclidean topology. Show that the component of each point $q \in \mathbb{Q}$ is $\{q\}$ and hence that $\mathbb{Q}$ is totally disconnected.

In Worked problem 1.2, we saw that this space is disconnected.

Solution

The proof is by contradiction. Let $q \in \mathbb{Q}$ and assume that C_q consists of more than one point, so there is $p \neq q$ with $p \in C_q$. Then there is an irrational number x such that $\min\{p, q\} < x < \max\{p, q\}$. Set

$U = \{r \in C_q : r < x\}$ and $V = \{r \in C_q : r > x\}$.

Since $\mathbb{Q}$ consists only of rational points, $x \notin C_q$.

Then U and V are non-empty and disjoint, and their union is C_q. Now $(-\infty, x) \cap \mathbb{Q}$ is open for the subspace topology on $\mathbb{Q}$, and so

$U = ((-\infty, x) \cap \mathbb{Q}) \cap C_q$

is open for the subspace topology on C_q. Similarly, V is open for the subspace topology on C_q. Thus $\{U, V\}$ is a disconnection of C_q. But this is impossible, since C_q is connected.

This contradiction shows that there can be no point $p \neq q$ in C_q. Thus, $C_q = \{q\}$ and so $\mathbb{Q}$ is totally disconnected. ∎

Problem 2.3

Show that any set X with at least two elements and carrying its discrete topology is totally disconnected.

You may have noticed that, in all the examples you have seen so far, the set X is the union of its mutually disjoint components. Thus, in Problem 2.2, $X = \{a\} \cup \{b, c\}$, in Worked problem 2.1, $\mathbb{Q} = \bigcup_{q \in \mathbb{Q}} \{q\}$ and, in Problem 2.3, $X = \bigcup_{x \in X} \{x\}$. We now show that this is true in general.

Theorem 2.6

Let $(X, \mathcal{T})$ be a topological space. Let us write $x \sim y$ to mean that $y \in C_x$. Then $\sim$ is an equivalence relation on X.

Consequently, X is the union of its mutually disjoint components.

Proof In order to prove that $\sim$ is an equivalence relation we must show that $\sim$ is:

(a) reflexive: for $x \in X$, $x \sim x$;
(b) symmetric: for $x, y \in X$, $x \sim y$ implies $y \sim x$;
(c) transitive: for $x, y, z \in X$, if $x \sim y$ and $y \sim z$, then $x \sim z$.

Reflexive Since $x \in C_x$, it follows immediately that $x \sim x$ for all $x \in X$.

Symmetric Suppose $x \sim y$ — that is, $y \in C_x$. Then there is a connected subset A of X containing both x and y. This implies that $A \subseteq C_y$, and so $x \in C_y$. Thus $y \sim x$.

Transitive Suppose $x \sim y$ and $y \sim z$. Let A be a connected subset of X containing x and y, and B be a connected subset of X containing y and z. Note that $A \cap B$ contains y, and so is non-empty. Therefore, by Theorem 1.5, $A \cup B$ is connected. Since x and z are in $A \cup B$, it follows that $z \in C_x$, and so $x \sim z$.

Hence $\sim$ is an equivalence relation, and so X can be written as the disjoint union of the equivalence classes defined by $\sim$. But, by definition of $\sim$, the equivalence class of a point x consists of all the points in C_x — namely C_x itself — and so each equivalence class is a component of X. ■

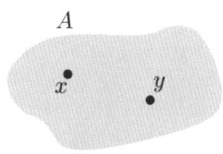

Figure 2.3 $x \sim y \Rightarrow y \sim x$

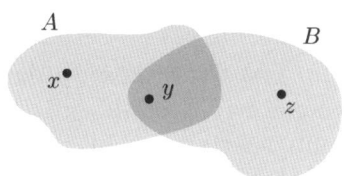

Figure 2.4 $x \sim y$, $y \sim z$ $\Rightarrow x \sim z$

Remark

One useful immediate consequence of this theorem, which was hinted at in the solution to Problem 2.2, is that, for each $y \in C_x$, $C_y = C_x$. In other words, the component of each point in C_x is C_x itself.

We end this section with the following result which states that each component of a set X is *closed*.

Lemma 2.7

Let $(X, \mathcal{T})$ be a topological space, and let $x \in X$. Then C_x, the component of x in $(X, \mathcal{T})$, is closed.

Problem 2.4

Prove Lemma 2.7.

Hint You may find Theorem 2.4 useful.

Problem 2.5

Determine the components of $X = \{a, b, c, d\}$ for the topology

$$\mathcal{T} = \{\varnothing, \{b\}, \{c\}, \{a, b\}, \{b, c\}, \{c, d\}, \{a, b, c\}, \{b, c, d\}, X\}.$$

This result, when combined with Theorem 2.6, tells us that a topological space can be split up into disjoint closed pieces, each of which is connected. These results are very useful when it comes to identifying the components of a set.

3 Connectedness in Euclidean spaces

After working through this section, you should be able to:
▶ show that $(\mathbb{R}^n, \mathcal{T}(d^{(n)}))$ is connected, for each $n \in \mathbb{N}$;
▶ specify which subsets of $\mathbb{R}$ are connected;
▶ appreciate the generalization of the Intermediate Value Theorem;
▶ classify intervals in $\mathbb{R}$;
▶ specify some connected subsets of $\mathbb{R}^n$.

In this section we carry out an extended study of the connected sets for one particular group of topological spaces: we find *all* the connected subsets of $\mathbb{R}$ with respect to the Euclidean topology, and show that the Euclidean space $\mathbb{R}^n$ is connected, for each n.

This enables us to prove a generalization of the Intermediate Value Theorem that we gave in *Unit A1*. This is an important result, and is one of the reasons why we investigate connectedness in topological spaces.

3.1 Connectedness in the real line

Our first objective is to show that $\mathbb{R}$ with the Euclidean topology is itself connected. Even though we expect this to be so intuitively, the proof is quite subtle. The difficulties arise from the fact that $\mathbb{Q}$ (with the subspace topology) is totally disconnected and yet possesses many of the properties of the real line. Thus our proof that $\mathbb{R}$ is connected has to rely on properties of $\mathbb{R}$ that $\mathbb{Q}$ does not possess. The required crucial property of $\mathbb{R}$ (that $\mathbb{Q}$ does not have) is the *least upper bound property*. This states that if $A \subseteq \mathbb{R}$ is a non-empty set that is bounded above, then A has a **least upper bound**, or **supremum** — a real number M such that:

(a) $a \leq M$, for all $a \in A$;
(b) if $M' < M$, then there is some $a \in A$ with $M' < a$.

We often write $\sup A = M$.
M is an upper bound of A.
M is the *least* upper bound.

For example, $\{x \in \mathbb{R} : x^2 < 2\}$ has a least upper bound of $\sqrt{2}$ in $\mathbb{R}$, while $\{x \in \mathbb{Q} : x^2 < 2\}$ does not have a least upper bound in $\mathbb{Q}$ (since $\sqrt{2} \notin \mathbb{Q}$).

Problem 3.1

Use the least upper bound property to show that if a non-empty set $A \subseteq \mathbb{R}$ is bounded below, then it has a **greatest lower bound**, or **infimum** — a real number m such that:

(a) $a \geq m$, for all $a \in A$,
(b) if $m' > m$, then there is some $a \in A$ with $a < m'$.

Hint Consider the set $B = \{-a : a \in A\}$.

We often write $\inf A = m$.
m is a lower bound of A.
m is the *greatest* lower bound.

> **Theorem 3.1**
>
> $\mathbb{R}$ with the Euclidean topology is connected.

Proof The proof is by contradiction. Suppose that $\mathbb{R}$ is disconnected, and let $\{U, V\}$ be a disconnection. Thus U and V are disjoint non-empty clopen subsets of $\mathbb{R}$ with union $\mathbb{R}$.

Consider the pair of intervals $(-\infty, 0]$ and $[0, \infty)$. They are both closed and their union is $\mathbb{R}$, but they are not disjoint — they have the point 0 in common.

Since $U \cup V = \mathbb{R}$, the point 0 lies in one of the sets U and V. Without loss of generality, we can suppose that $0 \in V$. Since $U \neq \varnothing$, at least one of $(-\infty, 0] \cap U$ and $[0, \infty) \cap U$ is non-empty.

Case 1: $(-\infty, 0] \cap U \neq \varnothing$

Let $A = (-\infty, 0] \cap U$.

Since A is the intersection of two closed sets, it is closed. It is also bounded above by 0.

> U is both closed and open.

Since $0 \in V$ and $U \cap V = \varnothing$, $0 \notin U$, and so 0 does not belong to A. Hence
$$A = (-\infty, 0) \cap U.$$

> The observation that removing 0 does not change A is the key to the proof.

Since A is the intersection of two open sets, it is open.

Hence A is clopen, non-empty and bounded above (by 0).

By the least upper bound property A has a least upper bound — denote this by M. If $M' < M$, then there is some $a \in A$ with $M' < a \leq M$, and so each neighbourhood of M intersects A. Therefore M is a closure point of A. Hence, since A is closed, $M \in A$.

> Each neighbourhood of M in $\mathbb{R}$ must contain an interval (M', M'') where $M' < M$ and $M'' > M$.

Since A is open, it contains an open interval centred at each of its points. So there is a number $r > 0$ such that $(M - r, M + r) \subseteq A$.

But then M is not the least upper bound of A, since A contains points y with $M < y < M + r$ — for example, $M + \frac{1}{2}r$.

We have arrived at a contradiction: so Case 1 does not apply.

Case 2: $[0, \infty) \cap U \neq \varnothing$

In this case, we let $A = [0, \infty) \cap U$, and use the greatest lower bound property to obtain a contradiction in a similar way: so Case 2 does not apply.

We conclude that $\mathbb{R}$ is connected. ∎

Intuition suggests that every interval is connected. We now show that this is indeed true, starting with the interval $(0, 1)$.

Worked problem 3.1 Show that the interval $(0, 1)$ is a connected subset of $\mathbb{R}$.

Solution

Observe that $\phi(x) = \frac{1}{2} + \frac{1}{\pi} \tan^{-1}(x)$ is a composite of basic continuous functions on $\mathbb{R}$ and so is continuous. Also $\phi(\mathbb{R}) = (0, 1)$. Since $\mathbb{R}$ is connected, Theorem 2.2 implies that $(0, 1)$ is connected. ∎

> See *Unit A1*.

Problem 3.2

Let $a < b$ be real numbers. Show that the interval (a, b) is a connected subset of $\mathbb{R}$.

Hint Find a suitable continuous map $\phi: (0, 1) \to (a, b)$.

Problem 3.3

Let $a < b$ be real numbers. Use the result of Problem 3.2 to show that the intervals $[a, b)$, $(a, b]$ and $[a, b]$ are connected.

We have now shown that all bounded intervals are connected. Similarly, by considering the map $\phi: \mathbb{R} \to (a, \infty)$ given by $\phi(x) = a + e^x$, we can show that (a, ∞) is connected for each $a \in \mathbb{R}$. Likewise we can use the map $\psi(x) = a - e^x$ to show that $(-\infty, a)$ is connected. It is then straightforward to use Theorem 2.4 to show that $(-\infty, a]$ and $[a, \infty)$ are connected.

We can now identify many subsets of $\mathbb{R}$ that are connected. Certainly $\emptyset$ and single-point subsets of $\mathbb{R}$ are connected. These sets can be interpreted as being degenerate intervals, namely $\emptyset = (a, a)$ and $\{a\} = [a, a]$.

See Problems 1.2 and 1.4.

We have thus shown that all the intervals in $\mathbb{R}$ are connected. In fact, the *only* connected subsets of $\mathbb{R}$ are intervals. In order to prove this, we need a characterization of intervals that distinguishes them from all other subsets of $\mathbb{R}$. This characterization depends on the fact that there are no 'holes' in an interval (in $\mathbb{R}$), and so if two points are in an interval, and a third point lies between them, then that third point must also belong to the interval. More precisely:

> a set $A \subseteq \mathbb{R}$ is an *interval* if and only if, whenever $a, b \in A$ and $c \in \mathbb{R}$ with $a < c < b$, then $c \in A$.

You can easily check that this characterization applies to all intervals of the form (a, b), $[a, b)$, $(a, b]$, $[a, b]$, $(-\infty, a)$, $(-\infty, a]$, $[a, \infty)$ and (a, ∞). It applies also to $(-\infty, \infty)$, which is the representation of $\mathbb{R}$ as an interval. It also trivially applies to the degenerate intervals $(a, a) = \emptyset$ and $[a, a] = \{a\}$. We now use this characterization to prove the promised result.

Theorem 3.2

A subset A of $\mathbb{R}$ is connected if and only if it is an interval.

Proof We have already shown that all intervals are connected. It remains only to show that if A is connected, then it is an interval. The proof is by contradiction.

Let A be connected, and suppose that A is *not* an interval.

Since A is not an interval, there must be real numbers a, b, c with $a, b \in A$ and $a < c < b$ such that $c \notin A$ — otherwise A would be an interval (by our characterization above). Let $\tilde{U} = (-\infty, c)$ and $\tilde{V} = (c, \infty)$: we claim that $U = \tilde{U} \cap A$ and $V = \tilde{V} \cap A$ form a disconnection of A. We leave the proof as a problem (Problem 3.4).

Since the empty set and single-point sets are intervals, A must contain at least two points. We do not require that A is bounded above or below.

This contradicts the fact that A is connected. We conclude that a connected subset of $\mathbb{R}$ must be an interval. ∎

Problem 3.4

Prove that the sets U and V in the above proof form a disconnection of A.

It is now a short step to the promised generalization of the *Intermediate Value Theorem* from *Unit A1*.

> **Theorem 3.3 Intermediate Value Theorem**
>
> Let $(X, \mathcal{T})$ be a connected topological space, and let $f: X \to \mathbb{R}$ be a $(\mathcal{T}, \mathcal{T}(d^{(1)}))$-continuous function. Let a and b be points of X. Then f takes each value between $f(a)$ and $f(b)$.

Proof Since f is continuous and X is connected, Theorem 2.2 implies that $f(X)$ is a connected subspace of $\mathbb{R}$, and so is an interval by Theorem 3.2. This interval contains $f(a)$ and $f(b)$, hence $f(X)$ contains each value between $f(a)$ and $f(b)$. ∎

Remark

This remarkable theorem is seen to be an immediate consequence of three things: the domain is connected, the only connected subsets of $\mathbb{R}$ are the intervals, and continuous images of connected sets are connected. Hence it is essentially a theorem about connectedness.

Problem 3.5

Let $f: \mathbb{R}^2 \to \mathbb{R}$ be the polynomial defined by

$$f(x_1, x_2) = 5x_1^5 + 5x_1^3 x_2^2 - x_2^5.$$

Show that there is a point in the plane at which f attains the value 42. (You may assume that the Euclidean space $(\mathbb{R}^2, \mathcal{T}(d^{(2)}))$ is connected and that the function f is $(\mathcal{T}(d^{(2)}), \mathcal{T}(d^{(1)}))$-continuous.)

The connectedness of $(\mathbb{R}^2, \mathcal{T}(d^{(2)}))$ will be proved in the next section.

3.2 Homeomorphic subsets of the real line

We now investigate which subsets of $\mathbb{R}$ are homeomorphic to each other. We know from Corollary 2.3 that connectedness is a topological invariant: that is, if two spaces are homeomorphic and one is connected, then so is the other. So $\mathbb{R}$ cannot be homeomorphic to any subset of $\mathbb{R}$ that is not connected.

It follows from Theorem 3.2 that any subset of $\mathbb{R}$ that is not an interval must be disconnected: for example, the set $\mathbb{R} - \{a\}$ is disconnected for each $a \in \mathbb{R}$.

Problem 3.6

Write down a disconnection of $\mathbb{R} - \{a\}$, for each $a \in \mathbb{R}$.

Since $\mathbb{R}$ is connected and $\mathbb{R} - \{a\}$ is disconnected, we deduce the following result.

> **Corollary 3.4**
>
> $\mathbb{R} - \{a\}$ is not homeomorphic to $\mathbb{R}$, for each $a \in \mathbb{R}$.

We know that every interval in $\mathbb{R}$ is connected. Is it possible that any two intervals are homeomorphic to each other?

We know that the function defined by $f(x) = a + (b-a)x$ is a homeomorphism between the open interval (a,b) and $(0,1)$ for $a < b$. Hence any two open intervals (a,b) and (c,d) are homeomorphic to $(0,1)$, and therefore to each other. Similarly, it can be shown that any two closed intervals $[a,b]$ and $[c,d]$ are homeomorphic, that $[a,b)$ and $[c,d)$ are homeomorphic and that $(a,b]$ and $(c,d]$ are homeomorphic. Furthermore, $(a,b]$ is homeomorphic to $[a,b)$, a suitable homeomorphism being the map defined by $f(x) = (a+b) - x$.

See *Unit A3*, Worked Problem 4.2.

So, bounded intervals of similar type are homeomorphic. What about intervals of different type? For example, can open intervals be homeomorphic to closed intervals? To answer these questions, we use the Restriction Rule for homeomorphisms, which states that if $(X, \mathcal{T}_X)$ and $(Y, \mathcal{T}_Y)$ are topological spaces, if $\phi\colon X \to Y$ is a homeomorphism and if $A \subseteq X$, then $\phi|_A\colon A \to \phi(A)$ is a homeomorphism between A and $\phi(A)$ with the subspace topologies inherited from $(X, \mathcal{T}_X)$ and $(Y, \mathcal{T}_Y)$.

Unit A3, Theorem 4.8.

Worked problem 3.2 Show that $A = (0,1)$ is not homeomorphic to $B = (0,1]$.

Solution

The proof is by contradiction. We know that A and B are both connected. Suppose that there is a homeomorphism $\phi\colon A \to B$.

Let $a \in A$ be the point for which $\phi(a) = 1$. Then, by the Restriction Rule for homeomorphisms, $A - \{a\}$ is homeomorphic to $\phi(A - \{a\}) = B - \{1\} = A$. This, however, is a contradiction, since $A - \{a\}$ is disconnected while A is connected. Thus A is not homeomorphic to B. ∎

A disconnection is $\{(0,a), (a,1)\}$.

It follows from Worked problem 3.2 that, in general, (a,b) is not homeomorphic to $(a,b]$, and hence (a,b) is not homeomorphic to $[a,b)$.

Problem 3.7

Show that $A = (0,1]$ is not homeomorphic to $B = [0,1]$.

It follows from Problem 3.7 that, in general, $[a,b]$ is not homeomorphic to $(a,b]$, and hence $[a,b]$ is not homeomorphic to $[a,b)$.

Similar arguments can be used to show that (a,b) is not homeomorphic to $[a,b]$.

We thus have five classes of intervals:
- the empty set $\varnothing$;
- the singleton sets $\{a\} = [a,a]$;
- the open intervals (a,b) for $a < b$ and (a,∞) and $(-\infty,a)$ and $\mathbb{R}$;
- the closed intervals $[a,b]$ for $a < b$;
- the intervals of the form $[a,b)$ and $(a,b]$ for $a < b$ and $[a,\infty)$ and $(-\infty,a]$.

Intervals in the same class are homeomorphic and intervals in different classes are not homeomorphic.

3.3 Products of connected spaces

Now that we know that the Euclidean line $\mathbb{R}$ is connected, we study the connectedness of $\mathbb{R}^n$, with the Euclidean topology, for $n > 1$.

We saw in *Units A2* and *A3* that many of the properties of $\mathbb{R}^n$ are immediate consequences of the fact that $\mathbb{R}^n$ is formed by taking an appropriate product of $\mathbb{R}$ with itself. In particular, the Euclidean topology on $\mathbb{R}^2$ is the product topology of the Euclidean topology on $\mathbb{R}$ with itself.

It seems plausible that connectedness should be preserved when we take products — for example, the product of an interval with an interval gives a rectangle in the plane (Figure 3.1), a set that we expect to be connected.

Figure 3.1

To show that $\mathbb{R}^2$ (or a rectangle) is connected, it is sufficient to show that the product of two connected topological spaces is itself a connected topological space.

Once we have shown that the product of two connected spaces is connected, we can use induction to deduce that n-fold products of connected spaces are also connected: for example, $\mathbb{R}^n$ is connected for each $n \in \mathbb{N}$.

Theorem 3.5

Let $(X, \mathcal{T}_X)$ and $(Y, \mathcal{T}_Y)$ be non-empty topological spaces. The product space $(X \times Y, \mathcal{T}_X \times \mathcal{T}_Y)$ is connected if and only if $(X, \mathcal{T}_X)$ and $(Y, \mathcal{T}_Y)$ are both connected.

Proof We prove that if X and Y are connected, then $X \times Y$ is connected. We leave the converse as a problem for you (Problem 3.8).

Suppose that $(X, \mathcal{T}_X)$ and $(Y, \mathcal{T}_Y)$ are connected. We must show that $(X \times Y, \mathcal{T}_X \times \mathcal{T}_Y)$ is also connected. To do this, it is enough to show that, for a given $z \in X \times Y$, the component of z is the whole space $X \times Y$.

So suppose that $z = (x, y) \in X \times Y$, and let C_z denote the component of z in $X \times Y$. In order to show that $C_z = X \times Y$, we must show that if $z' \in X \times Y$, then $z' \in C_z$.

So suppose that $z' = (x', y') \in X \times Y$. We show that $z' \in C_z$ by finding a connected subset A of $X \times Y$ that contains z and z'. We define $A = (X \times \{y\}) \cup (\{x'\} \times Y)$, as illustrated in Figure 3.2.

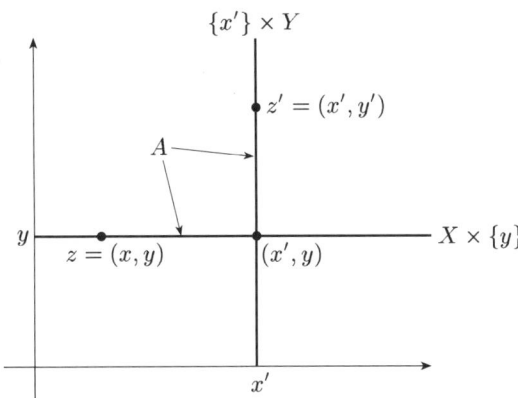

Figure 3.2 $A = (X \times \{y\}) \cup (\{x'\} \times Y)$

Clearly, both z and z' are elements of A, so it remains only to show that A is connected.

Now $X \times \{y\}$ (with its subspace topology) is a continuous image of X and $\{x'\} \times Y$ is a continuous image of Y. Thus, by Theorem 2.2, both $X \times \{y\}$ and $\{x'\} \times Y$ are connected.

We do not include the proofs of these technical results.

However, the point (x', y) is in both $X \times \{y\}$ and $\{x'\} \times Y$. Thus, by Theorem 1.5, A is connected. ∎

Problem 3.8

Prove the second part of this theorem, that if $(X, \mathcal{T}_X)$ and $(Y, \mathcal{T}_Y)$ are non-empty topological spaces and $(X \times Y, \mathcal{T}_X \times \mathcal{T}_Y)$ is connected, then so are $(X, \mathcal{T}_X)$ and $(Y, \mathcal{T}_Y)$.

Hint The projection functions are continuous.

Unit A3, Theorem 5.6.

Now that we know that the product of connected spaces is connected, and that $\mathbb{R}$ is connected, we can deduce that $\mathbb{R}^2$ is connected — and by induction, that $\mathbb{R}^n$ is connected for all $n > 1$. Similarly, it follows that the products of intervals in $\mathbb{R}$ are connected.

We are using the Euclidean topologies here.

Corollary 3.6

The Euclidean space $\mathbb{R}^n$ is connected. The n-dimensional 'rectangles' $I_1 \times \cdots \times I_n$ are connected, where I_i is an interval in $\mathbb{R}$.

Finally, we consider the connectedness properties of various sets in $\mathbb{R}^n$ of special interest. Our discussion is informal, as we omit the details of the homeomorphisms, though they are not especially difficult to write down.

The unit circle S is a connected subset of the plane: to see this, note that the mapping $t \mapsto (\cos t, \sin t)$ is a continuous mapping from $[0, 2\pi)$ onto S. By taking the topological product of S with itself (see Figure 3.3), we deduce that the torus is connected.

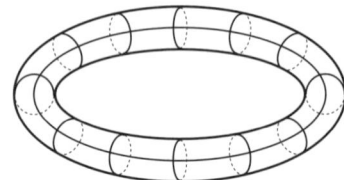

Figure 3.3

The punctured circle is homeomorphic to the Euclidean line $\mathbb{R}$, by means of stereographic projection (see Figure 3.4). Similarly, the punctured sphere is homeomorphic to the Euclidean plane (see Figure 3.5). Since the Euclidean plane is connected, so is the punctured sphere. The puncture is repaired by the operation of *closure*, and as closure preserves connection, by Theorem 2.4, the sphere is connected.

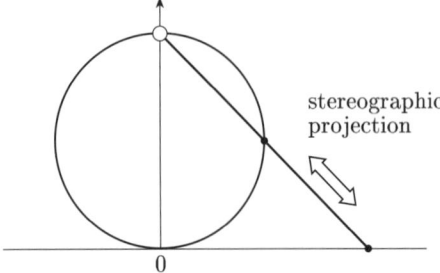

Figure 3.4

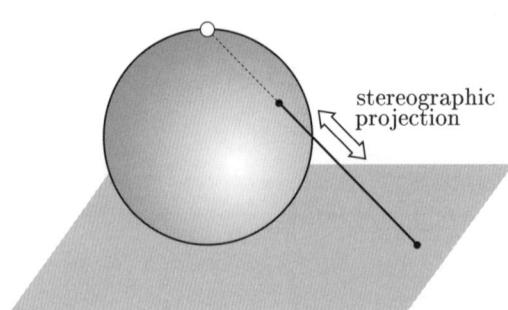

Figure 3.5

4 Path-connected spaces

After working through this section, you should be able to:
- explain what are meant by a *path* and a *path-connected* space;
- show that certain spaces are path-connected;
- state and make use of properties of path-connected spaces;
- appreciate the links and differences between path-connected and connected spaces.

We now introduce a refinement of connectedness, known as *path-connectedness*.

The basic idea is geometrically intuitive: we say that a space is *path-connected* if it is possible to find a 'path' between any two of its points (Figure 4.1). It turns out (as the name suggests) that path-connected spaces are always connected, and so one possible method for showing that a space is connected is to show that it is path-connected. However, not all connected spaces are path-connected.

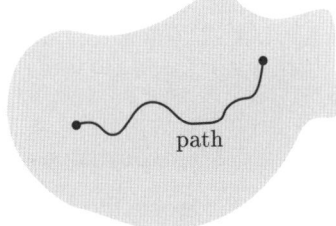

Figure 4.1

Section 5 discusses such a space in detail.

We cannot proceed further until we define what we mean by a *path* in a topological space. This concept is of wide interest in the applications of topology, even when topological language is not used. For example, an important construction in mathematics is the *path integral*. Path integrals are *not* part of this course, but if you have studied a course in complex analysis (contour integration), fluid mechanics (vortices), electromagnetism (the integral laws relating currents and magnetic fields), or differential geometry and its applications to general relativity, you have met path (or line) integrals, and hence paths.

We met this idea briefly in *Unit B1*.

4.1 What is a path?

What should we mean by a path? Imagine a snail moving on some surface in time (Figure 4.2). We imagine that as the snail moves, it leaves a trail behind it on the surface. Suppose the motion takes place from time $t=0$ to $t=1$, after which we take the snail off the surface and examine the trail it left. What will not be apparent is how the trail was made: Did the snail speed up, or slow down? Did it stop anywhere? Did it reverse back along a portion of the trail, then reverse again and go on its way? There are infinitely many ways that a given trail can be laid down, so the concept of a trail is inadequate for mathematical purposes.

If you find this analogy too unpleasant, think, instead, of the nib of a pen being drawn across the surface!

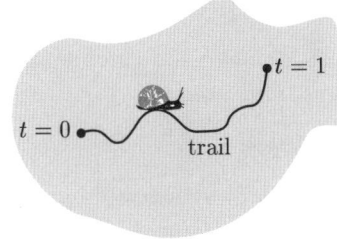

A better way to describe the snail's progress is to record where it is at each instant — that is, we consider its *path* to be the *function* that records the progress as a function of time. Mathematically, it is a function $p:[0,1] \to X$, where X is the surface the snail moves on. The interpretation is that the snail is at the point $p(t) \in X$ at time $t \in [0,1]$: in particular, it starts at $p(0)$ and finishes at $p(1)$. In this language, the trail is the image set $p([0,1]) \subseteq X$.

Figure 4.2

The only condition we demand of a path is that it be a *continuous function* when $[0,1]$ and the surface are given their Euclidean topologies: the snail may speed up, slow down, stop, reverse, change direction, as often and as abruptly as we please, but it must not 'break' its motion and jump.

The snail is not a grasshopper!

The trail may be extremely complicated and hard to visualize. It is known, for example, that there are paths on the plane whose trails completely fill the unit square! Such functions turn out to be nowhere differentiable, but this will not cause us any difficulty.

There is no mathematical reason as to why we should confine ourselves to paths on surfaces. All that we require is continuity, and so everything we have said so far holds if the codomain is an arbitrary topological space.

Definition

Let $(X, \mathcal{T})$ be a topological space and let $[0, 1]$ have the Euclidean subspace topology.

A function $p: [0, 1] \to X$ is a **path** in X if p is a $(\mathcal{T}(d^{(1)}), \mathcal{T})$-continuous function.

A path p **goes from** a **to** b in X if p is a path in X and if there are $t_x, t_y \in [0, 1]$ with $t_x \leq t_y$, $p(t_x) = a$ and $p(t_y) = b$. If the order in which the points are reached is unimportant, then we say that p **joins** a and b.

The **initial point** of a path p is the point $p(0)$ in X and the **final point** of p is the point $p(1)$.

The path is a **closed path** if the initial and final points of a path are the same (that is, $p(0) = p(1)$).

The **trail** of a path is its image set $p([0, 1])$.

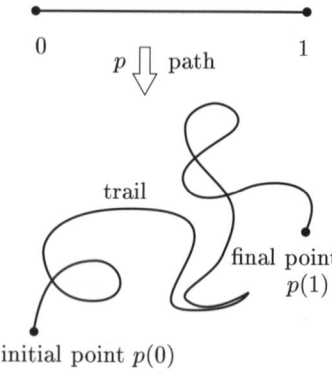

Figure 4.3

Remarks

(i) Since $[0, 1]$ with the Euclidean topology is connected, Theorem 2.2 tells us that the trail $p([0, 1])$ of a path p is a $\mathcal{T}$-connected subset of X.

(ii) The meaning of 'closed' in 'closed path' is different from its meaning in 'closed set'. This unfortunate coincidence of terminology is standard. It is usually clear from the context which meaning of 'closed' is intended.

(iii) For each point x in a topological space X, there is always a path joining a point x to itself — namely, the constant function $p: [0, 1] \to X$ given by $p(t) = x$.

In *Unit C2*, we show that if $(X, \mathcal{T})$ is metrizable, then the trail of p is a $\mathcal{T}$-closed set in X.

We saw in *Unit A3*, Theorem 4.1, that constant functions are continuous.

Worked problem 4.1

Let $\mathbb{R}^2$ have the Euclidean topology. Find a path with initial point $(0, 0)$ and final point $(1, 1)$.

Solution

The simplest example of such a path is probably the function $p: [0, 1] \to \mathbb{R}^2$ given by $p(t) = (t, t)$, illustrated in Figure 4.4. We need to verify that this p is continuous. In order to do so, we note that $p_1 \circ p$ and $p_2 \circ p$ are both continuous functions from $[0, 1]$ to $\mathbb{R}$, where p_1 and p_2 denote the usual projection functions from $\mathbb{R}^2$ to $\mathbb{R}$. Hence, by Theorem 5.7 of *Unit A3*, p is continuous. Since $p(0) = (0, 0)$ and $p(1) = (1, 1)$, p is a path with initial point $(0, 0)$ and final point $(1, 1)$. ∎

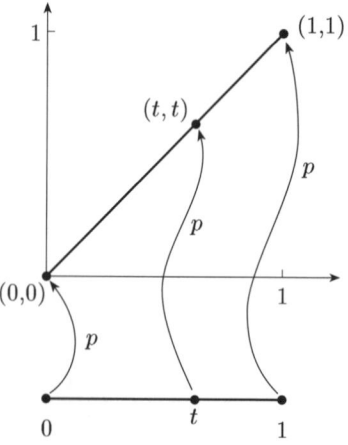

Figure 4.4

Problem 4.1

Let $X = [0,1] \times [0,1]$ with the subspace topology inherited from the Euclidean plane. Show that there is a path joining $(0,0)$ and $(1, \frac{1}{2})$ in X.

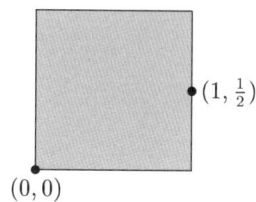

Figure 4.5 $X = [0,1] \times [0,1]$

We next show that paths can be joined.

Lemma 4.1

Let $(X, \mathcal{T})$ be a topological space and let $a, b, c \in X$. Suppose that there is a path p_1 in X with initial point a and final point b, and a path p_2 in X with initial point b and final point c. Then there is a path that joins a and c in X.

Proof The basic idea here is simple: we define a path joining a and c by doing p_1 followed by p_2. There is a problem, however: both p_1 and p_2 are defined for $t \in [0,1]$, and what we would like is to trace the whole of p_1 for $t \in [0, \frac{1}{2}]$ and then trace the whole of p_2 for $t \in [\frac{1}{2}, 1]$, as illustrated in Figure 4.6. The way to do this is to trace out both p_1 and p_2 at double their original 'speed'. Formally, we do this as follows.

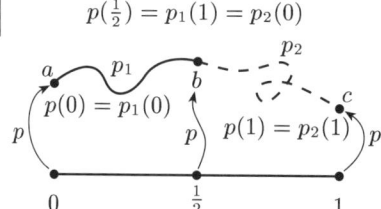

Figure 4.6

Define $p: [0,1] \to X$ by

$$p(t) = \begin{cases} p_1(2t) & \text{for } 0 \leq t \leq \frac{1}{2}, \\ p_2(2t-1) & \text{for } \frac{1}{2} \leq t \leq 1. \end{cases}$$

Certainly $p(0) = p_1(0) = a$ and $p(1) = p_2(1) = c$ and so it remains only to show that p is continuous. Note that $p(\frac{1}{2}) = p_1(1) = p_2(0) = b$.

Let U be an open set in X; we must show that $p^{-1}(U)$ is open in $[0,1]$ for the Euclidean topology. Now

$$\begin{aligned} p^{-1}(U) &= \{t \in [0,1] : p(t) \in U\} \\ &= \{t \in [0, \tfrac{1}{2}] : p(t) \in U\} \cup \{t \in [\tfrac{1}{2}, 1] : p(t) \in U\} \\ &= \{t \in [0, \tfrac{1}{2}] : p_1(2t) \in U\} \cup \{t \in [\tfrac{1}{2}, 1] : p_2(2t-1) \in U\} \\ &= \{r \in [0,1] : p_1(r) \in U\} \cup \{s \in [0,1] : p_2(s) \in U\} \\ &= p_1^{-1}(U) \cup p_2^{-1}(U). \end{aligned}$$

We have made the substitutions $r = 2t$ and $s = 2t - 1$.

Thus, since p_1 and p_2 are continuous, $p^{-1}(U)$ is an open set. Hence p is continuous, and so gives a path in X that joins a and c. ∎

In the next problem, we ask you to show that paths can be traced backwards.

Problem 4.2

Let $(X, \mathcal{T})$ be a topological space and let $a, b \in X$. Suppose that $p_1: [0,1] \to X$ is a path in X with initial point a and final point b. Show that there is a path p_2 in X with initial point b and final point a.

We end this subsection by looking at an example of a path in a set with a topology other than the Euclidean topology.

Worked problem 4.2

Let $X = \{a, b\}$ and let $\mathcal{T}$ be the topology $\{\varnothing, \{a\}, \{a, b\}\}$ on X. Show that the function $p\colon [0,1] \to X$ defined by

$$p(t) = \begin{cases} a & \text{if } t \in [0, \tfrac{1}{2}), \\ b & \text{if } t \in [\tfrac{1}{2}, 1], \end{cases}$$

is a path joining a and b.

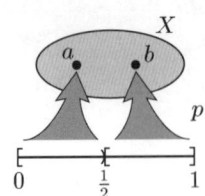

Figure 4.7

Solution

By definition, a and b are the images of points in $[0, 1]$. Thus to show that p is a path joining a and b it remains to show that p is continuous. To do this, we must show that the inverse image of each set in $\mathcal{T}$ is an open subset of $[0, 1]$.

Now $p^{-1}(\varnothing) = \varnothing$ and $p^{-1}(\{a, b\}) = [0, 1]$, both of which are open for the Euclidean subspace topology on $[0, 1]$. Also $p^{-1}(\{a\}) = [0, \tfrac{1}{2})$. Now $[0, \tfrac{1}{2}) = (-1, \tfrac{1}{2}) \cap [0, 1]$, and so $[0, \tfrac{1}{2})$ is open for the Euclidean subspace topology on $[0, 1]$.

Thus p is a path joining a and b. ∎

Note that had we chosen to attach the point $t = \tfrac{1}{2}$ to the point a, then the resulting function would not have been continuous.

4.2 Examples of path-connected spaces

Now that we have the notion of a path, we can define what we mean for a topological space to be path-connected.

Definition

Let $(X, \mathcal{T})$ be a topological space.

The space $(X, \mathcal{T})$ is **path-connected** if, for each a and b in X, there is a path in X that joins a and b.

A set $A \subseteq X$ is **path-connected** if $(A, \mathcal{T}_A)$ is path-connected.

$\mathcal{T}_A$ denotes the subspace topology on A.

Remarks

(i) If $X = \varnothing$, then X is certainly path-connected, since there are no points that need to be joined.

(ii) If $X = \{x\}$ has only one point, the constant function $f\colon [0, 1] \to X$ given by $f(t) = x$ is a path, and so X is path-connected.

(iii) When we wish to emphasize the topology, we use the term $\mathcal{T}$-*path-connected*.

Worked problem 4.3

Show that $\mathbb{R}$ with the Euclidean topology is path-connected.

Solution

Let $a, b \in \mathbb{R}$; we must find a path joining a and b in $\mathbb{R}$. We already know how to define such a path if $a = b$, so suppose that $a \neq b$. Without loss of generality we shall assume that $a < b$.

There are many such paths. The standard one is the straight-line path $P\colon [0, 1] \to \mathbb{R}$ defined by

$$P(t) = (1 - t)a + tb,$$

illustrated in Figure 4.8. We prove that P is a path from a to b.

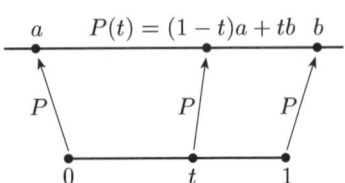

Figure 4.8

The linear function defined by $P(t)$ is a basic continuous function on $\mathbb{R}$. Hence $P\colon [0,1] \to \mathbb{R}$ is continuous as it is the restriction to $[0,1]$ of a basic continuous function on $\mathbb{R}$. Moreover, $P(0) = a$ and $P(1) = b$. Hence, P is a path in $\mathbb{R}$ from a to b. Thus $\mathbb{R}$ is path-connected. ∎

Problem 4.3

Show that $[a,b]$ with the Euclidean subspace topology is path-connected.

Remark

The result of Problem 4.3 enables us to deduce that every interval I in $\mathbb{R}$ is path-connected. We know that the empty interval is path-connected. If I is not empty, for each a and b in I ($a \leq b$) we know that $[a,b]$ is path-connected and hence there is a path from a to b in $[a,b] \subseteq I$. Thus I is path-connected.

The result in Worked problem 4.3 generalizes to $\mathbb{R}^n$ (for any $n \in \mathbb{N}$), as we ask you to show in the following problem.

Problem 4.4

Let $\mathbb{R}^n$ have the Euclidean topology, and, for $\mathbf{a}, \mathbf{b} \in \mathbb{R}^n$, define $P\colon [0,1] \to \mathbb{R}^n$ by

$$P(t) = (1-t)\mathbf{a} + t\mathbf{b}.$$

Prove that P is a path from $\mathbf{a}$ to $\mathbf{b}$, and hence that $\mathbb{R}^n$ is path-connected.

We can use the result of Problem 4.4 to show that every open ball in $\mathbb{R}^n$ is path-connected for the Euclidean topology. The first step is to show that the unit open ball in $\mathbb{R}^n$ is path-connected for the Euclidean topology.

Worked problem 4.4

Show that the unit open ball B in $\mathbb{R}^n$ with the Euclidean subspace topology is path-connected.

Solution

Let $\mathbf{a}, \mathbf{b} \in B$. We know from Problem 4.4 that $P(t) = (1-t)\mathbf{a} + t\mathbf{b}$ defines a path from $\mathbf{a}$ to $\mathbf{b}$ in $\mathbb{R}^n$. Thus, by Theorem 4.6 of *Unit A3*, we can deduce that P defines a path in B if we can show that $P(t)$ belongs to B for all $t \in [0,1]$ — that is,

$$d^{(n)}(P(t), \mathbf{0}) < 1.$$

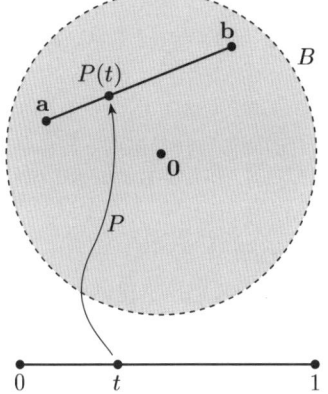

Figure 4.9

We have:

$$\begin{aligned} d^{(n)}(P(t), \mathbf{0}) &= d^{(n)}((1-t)\mathbf{a} + t\mathbf{b}, \mathbf{0}) && \text{(definition of } P\text{)} \\ &\leq d^{(n)}((1-t)\mathbf{a} + t\mathbf{b}, t\mathbf{b}) + d^{(n)}(t\mathbf{b}, \mathbf{0}) && \text{(Triangle Inequality)} \\ &= d^{(n)}((1-t)\mathbf{a}, \mathbf{0}) + d^{(n)}(t\mathbf{b}, \mathbf{0}) && \text{(translating by } t\mathbf{b}\text{)} \\ &= (1-t)d^{(n)}(\mathbf{a}, \mathbf{0}) + t d^{(n)}(\mathbf{b}, \mathbf{0}) && \text{(factorizing the scalar factors)} \\ &< (1-t) + t = 1. && (\mathbf{a}, \mathbf{b} \in B) \end{aligned}$$

Therefore B is path-connected. ∎

Since every open ball in $\mathbb{R}^n$ is homeomorphic to the unit open ball, if we can show that path-connectedness is a topological invariant, then we can deduce that all open balls are path-connected. A proof almost identical to that in Worked problem 4.4 allows us to show that the unit closed ball in $\mathbb{R}^n$ is path-connected, and thus to deduce that every closed ball in $\mathbb{R}^n$ is

We prove this in Theorem 4.4.

path-connected. So far, we have just considered path-connectedness for the Euclidean topology. We now consider a different topology on $\mathbb{R}$.

Worked problem 4.5

Show that $\mathbb{R}$ with the 0-deleted-point topology $\mathcal{T}_0$ is path-connected.

It follows from Problem 1.8 that this space is connected.

Solution

We use a method that is frequently useful when we wish to show that a particular space is path-connected: we fix a point a in the space and then show that we can connect it to any other point via a path. The result of Problem 4.2, together with Lemma 4.1, then imply that each pair of points in the space can be joined by a path (via a), as shown in Figure 4.10, and so the space is path-connected.

Figure 4.10

There is a point in $(\mathbb{R}, \mathcal{T}_0)$ that is special: the point 0. In view of the above, we need only show how to construct a path from 0 to any number x.

Remember that in the 0-deleted-point topology, the point 0 is not deleted from $\mathbb{R}$.

Let $x \in \mathbb{R}$. We find a path $p\colon [0,1] \to \mathbb{R}$ from 0 to x, with $p(0) = 0$ and $p(1) = x$. Define p by

$$p(t) = \begin{cases} 0 & \text{for } t = 0, \\ x & \text{for } 0 < t \leq 1. \end{cases}$$

Clearly, $p(0) = 0$ and $p(1) = x$. To show that p is continuous, we must show that the inverse image of every set in $\mathcal{T}_0$ is an open subset of $[0,1]$.

If U is an open subset of X, then either $U = X$ or $0 \notin U$.

If $U = X$, then $p^{-1}(U) = [0,1]$, an open subset of $[0,1]$ (with the Euclidean subspace topology).

If $0 \notin U$, then

$$p^{-1}(U) = \{t \in [0,1] : p(t) \in U\} = \begin{cases} \varnothing & \text{if } x \notin U, \\ (0,1] & \text{if } x \in U. \end{cases}$$

In either case, $p^{-1}(U)$ is an open subset of $[0,1]$ (with the Euclidean subspace topology). We conclude that p is a path joining 0 to x, and hence that $(\mathbb{R}, \mathcal{T}_0)$ is path-connected. ∎

Remark

Unfortunately, there is no algorithm for finding a path between two points of a given space; determining whether a space is path-connected ultimately requires skill and judgement.

Problem 4.5

Let X be a set containing at least two points, and let $\mathcal{T}$ be the indiscrete topology on X. Show that $(X, \mathcal{T})$ is path-connected.

$\mathcal{T} = \{\varnothing, X\}$. It follows from Problem 1.5 that this space is connected.

Hint Every function $f\colon [0,1] \to X$ is $(\mathcal{T}(d^{(1)}), \mathcal{T})$-continuous (see Theorem 1.2 of *Unit A3*).

You may have noticed that all the examples of path-connected spaces that you have met so far are also connected. This is no accident, as the following theorem shows.

Theorem 4.2

Let $(X, \mathcal{T})$ be a topological space. If X is path-connected, then X is connected.

The converse need not hold. A counter-example is given in Section 5.

Proof If X is empty, then there is nothing to prove, so we can suppose that $X \neq \varnothing$.

Fix some $a \in X$. If X is path-connected, then each point $x \in X$ can be joined to a by some path in X. Each such path defines a trail that (being the continuous image of the connected set $[0,1]$) is, by Theorem 2.2, a connected set, containing both a and x. The union of all these connected trails is the whole of X, and each such trail contains the point a. It follows from Theorem 1.5 that the union is a connected set. Thus X is connected. ∎

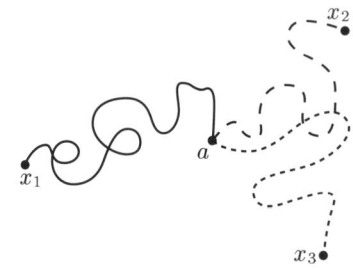

Figure 4.11

It follows from Theorem 4.2 that one way to show that a space is connected is to show that it is path-connected.

Worked problem 4.6

Let $A \subseteq \mathbb{R}^2$ be a countable set. Show that, for the Euclidean subspace topology, $\mathbb{R}^2 - A$ is path-connected, and hence connected.

Solution

We first observe that $\mathbb{R}^2 - A \neq \varnothing$, since $\mathbb{R}^2$ is uncountable and A is countable.

Let **a** and **b** be distinct points of $\mathbb{R}^2 - A$. We describe how to find a path joining **a** and **b**.

Let L be the perpendicular bisector of the line segment joining **a** and **b**. For each $\mathbf{x} \in L$, let $p_{\mathbf{x}} \colon [0,1] \to \mathbb{R}^2$ be a path joining **a** to **b** with trail as illustrated in Figure 4.12.

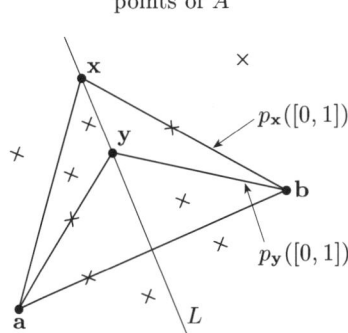

Figure 4.12

Let **x** and **y** be distinct points in L; then the trails of their corresponding paths $p_{\mathbf{x}}([0,1])$ and $p_{\mathbf{y}}([0,1])$ meet only at the points **a** and **b**. In particular, $A \cap p_{\mathbf{x}}([0,1])$ is disjoint from $A \cap p_{\mathbf{y}}([0,1])$, whenever $\mathbf{x} \neq \mathbf{y}$. Since L is uncountable and A is countable, this means that there is $\mathbf{x} \in L$ for which $p_{\mathbf{x}}([0,1]) \cap A = \varnothing$. For this **x**, $p_{\mathbf{x}}$ is a path in $\mathbb{R}^2 - A$ joining **a** to **b**.

Thus $\mathbb{R}^2 - A$ is path-connected. It follows from Theorem 4.2 that $\mathbb{R}^2 - A$ is connected. ∎

Remarks

(i) This argument works for any $\mathbb{R}^n$ with $n > 1$. However, it fails for $\mathbb{R}$ — removing just one point from $\mathbb{R}$ gives a disconnected set (Theorem 3.2). Using the invariance of path-connectedness under homeomorphisms (Theorem 4.4), we can deduce that $\mathbb{R}^n$ (for $n > 1$) is not homeomorphic to $\mathbb{R}$.

(ii) Since the empty set is countable, this result confirms that $\mathbb{R}^2$ is path-connected.

We ask you to prove this in one of the exercises for this unit.

Problem 4.6

Let X be a set containing at least two points and let $\mathcal{T}$ be the discrete topology on X. Show that X is not path-connected.

Hint Use the result of Worked problem 1.1.

We now use the result of Theorem 4.2 to show that $(C[0,1], \mathcal{T}(d_{\max}))$ is connected. Recall that $C[0,1]$ is the set of continuous functions on $[0,1]$ and that the max metric on $C[0,1]$ is defined by

$$d_{\max}(f,g) = \max\{|g(x) - f(x)| \colon x \in [0,1]\}.$$

Unit A2, Subsection 2.3.

Worked problem 4.7

Show that $(C[0,1], \mathcal{T}(d_{\max}))$ is path-connected, and hence connected.

Solution

Let f and g be functions in $C[0,1]$. Consider the function $P\colon [0,1] \to C[0,1]$ given by

$$P(t) = (1-t)f + tg,$$

so that, for $x \in [0,1]$ and $t \in [0,1]$,

$$P(t)(x) = (1-t)f(x) + tg(x).$$

We ask you to show that P is continuous, in Problem 4.7. Note that $P(0) = f$ and $P(1) = g$. We deduce that P is a path from f to g, and $C[0,1]$ is path-connected.

It now follows from Theorem 4.2 that $C[0,1]$ is connected. ∎

Problem 4.7

By bounding $d_{\max}(P(t), P(s))$ in terms of $|t - s|$ and $d_{\max}(f, g)$, prove that P is continuous.

4.3 Properties of path-connected spaces

We showed in Section 2 that connectedness is preserved under continuous mappings, and hence is a topological invariant. The same is true for path-connectedness.

Theorem 2.2 and Corollary 2.3.

Theorem 4.3

Let $(X, \mathcal{T}_X)$ and $(Y, \mathcal{T}_Y)$ be topological spaces, let X be path-connected, and let $f\colon X \to Y$ be $(\mathcal{T}_X, \mathcal{T}_Y)$-continuous. Then $f(X)$ is path-connected.

Proof Let $a, b \in f(X)$. By the definition of $f(X)$, there exist $c, d \in X$ such that $f(c) = a$ and $f(d) = b$. Since X is path-connected, there exists a path $p\colon [0,1] \to X$ from c to d. Since f and p are continuous, $f \circ p\colon [0,1] \to f(X)$ is a path from $f(c) = a$ to $f(d) = b$ in $f(X)$ (with the subspace topology inherited from $(Y, \mathcal{T}_Y)$). Thus $f(X)$ is path-connected. ∎

Remark

In our study of surfaces, we required that our surfaces be path-connected, and implicitly used the fact that, since polygons in the plane are path-connected, then so are any surfaces formed from them via edge identifications. This result follows directly from Theorem 4.3, since a surface formed by edge identifications from a polygon X with the subspace topology $\mathcal{T}$ is homeomorphic to the identification space $(I(X), \mathcal{T}_f)$ and $I(X)$ is a $(\mathcal{T}, \mathcal{T}_f)$-continuous image of X.

Unit B1, Subsection 1.5.

Unit B1, Subsection 4.1.

We can deduce from Theorem 4.3 that if $(X, \mathcal{T})$ is path-connected then any topological space *homeomorphic* to $(X, \mathcal{T})$ is also path-connected.

> **Theorem 4.4**
>
> Path-connectedness is a topological invariant.

We know from Worked problem 4.6 that $\mathbb{R}^2$ is path-connected, but it is natural to ask whether we could have proved this result by using the fact that $\mathbb{R}^2$ is the product of $\mathbb{R}$ with itself: do products preserve path-connectedness? The answer is yes.

> **Theorem 4.5**
>
> Let $(X, \mathcal{T}_X)$ and $(Y, \mathcal{T}_Y)$ be topological spaces. The product space $(X \times Y, \mathcal{T}_X \times \mathcal{T}_Y)$ is path-connected if and only if $(X, \mathcal{T}_X)$ and $(Y, \mathcal{T}_Y)$ are both path-connected.

The proof is similar to that given for Theorem 3.5, and we omit it.

As with connectedness, this product result tell us that, for any intervals $I_1, \ldots, I_n$ of $\mathbb{R}$, $I_1 \times \cdots \times I_n$ is path-connected. In particular, if $I_k = \mathbb{R}$ for each k, we have another proof that $\mathbb{R}^n$ is path-connected.

Once we know that a space is path-connected, we can deduce that, with any smaller topology, the resulting space is path-connected.

> **Theorem 4.6**
>
> Let $(X, \mathcal{T}_1)$ be a path-connected topological space, and let $\mathcal{T}_2 \subseteq \mathcal{T}_1$ be a coarser topology for X. Then $(X, \mathcal{T}_2)$ is path-connected.

Problem 4.8

Prove Theorem 4.6.

We now prove a result corresponding to Theorem 1.5 for connected spaces.

> **Theorem 4.7**
>
> Let $(X, \mathcal{T})$ be a topological space, and let $\{A_i : i \in I\}$ be a family of path-connected subsets of X whose intersection is non-empty. Then $A = \bigcup_{i \in I} A_i$ is path-connected.

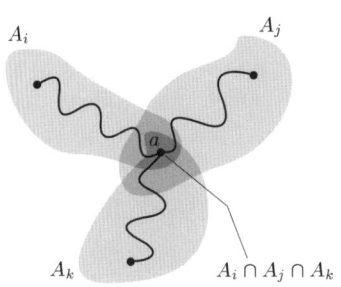

Figure 4.13

Proof Let $x, y \in A$; we seek a path in A joining x and y.

Since $\bigcap_{i \in I} A_i \neq \emptyset$, we can fix a point $a \in \bigcap_{i \in I} A_i$.

To show that A is path-connected, we first show that, for each $x \in A$, there is a path in A that joins x to a. Let $x \in A$. Since $x \in A$, there is $i \in I$ such that $x \in A_i$. Moreover, $a \in A_i$ and so, since A_i is path-connected, there is a path in A_i that joins x to a. We can now use Lemma 4.1 and the result of Problem 4.2 to find a path in A between any two points $x, y \in A$. ∎

Finally, we note that path-connectedness is not preserved under closure; in this respect it differs from connectedness. For an example, see the topologist's cosine in Section 5.

5 The topologist's cosine

After working through this section, you should be able to:
▶ describe an example of a topological space that is connected, but not path-connected.

This section presents an example of a topological space that is connected but not path-connected. One property of this example is that it is the closure of a path-connected set, and so it illustrates one way in which path-connected spaces differ from connected ones.

To obtain an example of such a set, and at the same time allow the possibility of some geometric visualization, we construct a subset of the plane with the required properties.

Our inspiration comes from the graph of the $\cos(\pi/x)$ function that we investigated in *Unit A1*. We defined this function $f: \mathbb{R} \to \mathbb{R}$ by

$$f(x) = \begin{cases} \cos(\pi/x) & \text{if } x \neq 0, \\ 0 & \text{if } x = 0. \end{cases}$$

Unit A1, Worked problem 2.3.

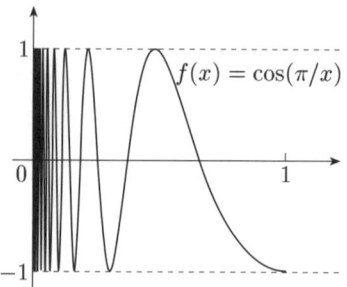

This function f is continuous everywhere except at 0. In fact, it is not possible to redefine this function at 0 to make it continuous (with respect to the Euclidean topology). If we consider the graph of the function f restricted to $(0, 1]$, we have the set

$$A = \{(x, \cos(\pi/x)) : 0 < x \leq 1\} \subset \mathbb{R}^2,$$

shown in Figure 5.1. It is plausible that this set is path-connected for its subspace topology inherited from the Euclidean topology for the plane.

Figure 5.1

What is the closure of this set in the plane? It seems likely that $\mathrm{Cl}(A)$ includes the vertical line segment $B = \{0\} \times [-1, 1]$ on the y-axis, shown in Figure 5.2. In fact, $\mathrm{Cl}(A) = A \cup B$. Is this set path-connected? The answer is no, and we can obtain an idea of why it isn't if we try to visualize how to construct a path between a point in B and a point in A. However, $\mathrm{Cl}(A)$ is connected.

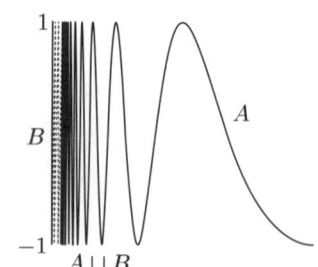

Our example, inspired by the $\cos(\pi/x)$ function, is constructed from line segments. This will make it much easier to determine whether a given element of the plane lies in the set.

Figure 5.2

Working out all the details in this example is a long process. Most of it is what pure mathematicians call *routine*, a word that should not be confused with *easy* or *elementary*. Routine work consists of providing details of standard pieces of analysis, such as proving the continuity of certain functions, or finding an upper bound for a set.

You can lighten your workload by omitting the details, and focusing simply on the gist of the argument.

We begin by constructing a straight-line version of the set A we have just described.

For $n \in \mathbb{N}$, let A_n denote the set in the plane consisting of the line segment joining the point $(\frac{1}{n+1}, 0)$ to the point $(\frac{1}{2}(\frac{1}{n} + \frac{1}{n+1}), 1)$, together with the line segment joining $(\frac{1}{2}(\frac{1}{n} + \frac{1}{n+1}), 1)$ to $(\frac{1}{n}, 0)$, as shown in Figure 5.3. The algebraic description is unpleasant, but for future reference we give it:

$$A_n = \left\{(s, 2n(n+1)s - 2n) : \tfrac{1}{n+1} \leq s \leq \tfrac{1}{2}\left(\tfrac{1}{n} + \tfrac{1}{n+1}\right)\right\}$$
$$\cup \left\{(s, 2(n+1) - 2n(n+1)s) : \tfrac{1}{2}\left(\tfrac{1}{n} + \tfrac{1}{n+1}\right) \leq s \leq \tfrac{1}{n}\right\}.$$

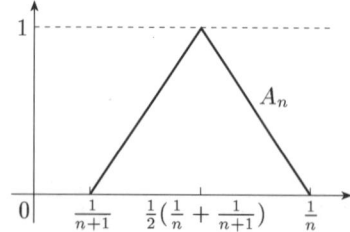

Figure 5.3

Let A be the plane subset
$$A = \bigcup_{n=1}^{\infty} A_n.$$
Then A is the *graph* of the function $l \colon (0,1] \to \mathbb{R}$ given by
$$l(x) = \begin{cases} 2n(n+1)x - 2n & \text{if } \frac{1}{n+1} < x \leq \frac{1}{2}\left(\frac{1}{n} + \frac{1}{n+1}\right), \ n \in \mathbb{N}, \\ 2(n+1) - 2n(n+1)x & \text{if } \frac{1}{2}\left(\frac{1}{n} + \frac{1}{n+1}\right) < x \leq \frac{1}{n}, \ n \in \mathbb{N}, \end{cases}$$
and shown in Figure 5.4. It is also the *image set* of the function
$$L \colon (0,1] \to \mathbb{R}^2 \text{ given by } L(x) = (x, l(x)).$$
Let B be the vertical line segment joining the origin to $(0,1)$:
$$B = \{(0, y) \colon 0 \leq y \leq 1\}.$$
The *topologist's cosine* is the plane set
$$C = A \cup B,$$
consisting of all the points of A and B (Figure 5.5).

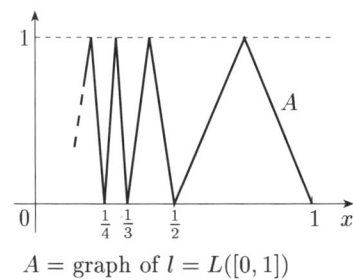

$A =$ graph of $l = L([0,1])$

Figure 5.4

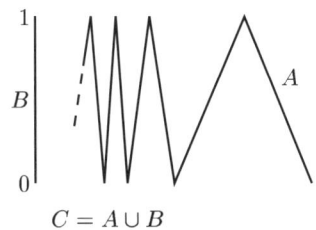

$C = A \cup B$

Figure 5.5

The topologies throughout are the appropriate Euclidean ones.

Lemma 5.1

The functions $l \colon (0,1] \to \mathbb{R}$ and $L \colon (0,1] \to \mathbb{R}^2$ are continuous.

Proof (sketch) We omit the routine proof that l is continuous.

Let p_1 and p_2 be the projection functions from $\mathbb{R}^2$ to $\mathbb{R}$. Then $(p_1 \circ L)(x) = x$ and $(p_2 \circ L)(x) = l(x)$, and so $p_1 \circ L$ and $p_2 \circ L$ are both continuous functions. Hence L is continuous, by Theorem 5.7 of *Unit A3*. ∎

This allows us to show that A is a path-connected (and hence connected) subset of the plane.

Corollary 5.2

A is path-connected for the Euclidean subspace topology.

Proof The set A is the image set of the path-connected set $(0,1]$ under the continuous function L. So, by Theorem 4.3, A is path-connected. ∎

We now show that C is a connected subset of the plane.

Theorem 5.3

C is connected for the Euclidean subspace topology.

Proof In order to show that C is connected, we let $\mathcal{T}_C$ denote the Euclidean subspace topology on C and show that the closure of A for $\mathcal{T}_C$ is C. Then, since A is connected (by Corollary 5.2 and Theorem 4.2), it follows from Theorem 2.4 that C is connected.

Since the whole space in question is C, it is certainly the case that $\mathrm{Cl}(A) \subseteq C$.

It remains only to show that each point in $C - A = B$ is a closure point of A in C.

Let $\mathbf{b} = (0, b) \in B$, so $0 \leq b \leq 1$. We show that $\mathbf{b}$ is a closure point of A by showing that each $\mathcal{T}_C$-neighbourhood of $\mathbf{b}$ meets A. Since we are working in $\mathbb{R}^2$, it suffices to show that each open ball with centre $\mathbf{b}$ meets C. Let U be an open ball of radius r centred at $\mathbf{b}$. We must find a point of A that lies in U. Figure 5.6 suggests that many such points exist.

A $\mathcal{T}_C$-neighbourhood of $\mathbf{b}$ is the intersection of a $\mathcal{T}(d^{(2)})$-neighbourhood of $\mathbf{b}$ with C, and the open balls in the plane form a base for $\mathcal{T}(d^{(2)})$ (*Unit A3*, Theorem 5.1).

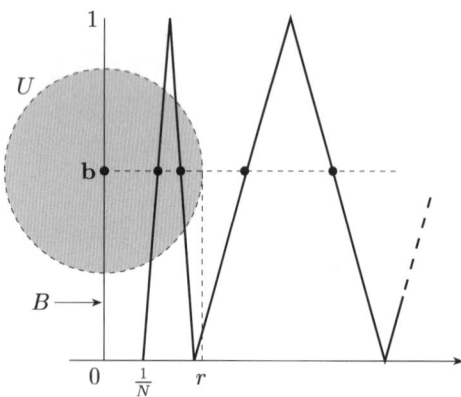

Figure 5.6

Consider A_n. There is at least one point in A_n with y-coordinate equal to b; for example, the point (x_n, b), where
$$x_n = \frac{b + 2n}{2n(n+1)}.$$
For, since $0 \leq b \leq 1$,
$$\frac{1}{n+1} \leq x_n \leq \frac{2n+1}{2n(n+1)} = \frac{1}{2}\left(\frac{1}{n} + \frac{1}{n+1}\right),$$
and so $l(x_n) = 2n(n+1)x_n - 2n = (b + 2n) - 2n = b$.

Thus, if $x_n < r$, then $(x_n, b) \in U \cap A$. To guarantee that $x_n < r$, we just need $n \geq N$, where $N \in \mathbb{N}$ is chosen so that
$$\frac{b + 2N}{2N(N+1)} < r.$$
That is, $2N((N+1)r - 1)) > b$; such an N always exists.

Thus, for $n > N$, $A_n \cap U \neq \varnothing$ and so $A \cap U \neq \varnothing$. Hence $\mathbf{b} \in \mathrm{Cl}(A)$. Thus $\mathrm{Cl}(A) = C$ and so C is connected. ∎

It remains to show that C is not path-connected.

Theorem 5.4

C is not path-connected for the Euclidean subspace topology.

Proof We have already observed in Corollary 5.2 that A is path-connected. Since B is also path-connected, the difficulty must lie in finding a path from a point in B to a point in A.

The proof is by contradiction. Suppose that C is path-connected: in particular, there is a continuous function $P\colon [0,1] \to C$ with initial point $P(0) = (0,0)$ in B and final point $P(1) = (1,0)$ in A, as shown in Figure 5.7.

To describe P, we use the language of the path of a particle as a function of time. At time t, the particle is at the point $P(t)$ in $\mathbb{R}^2$. Writing $P(t) = (P_1(t), P_2(t))$ introduces two new continuous functions, $P_1\colon [0,1] \to [0,1]$ and $P_2\colon [0,1] \to [0,1]$. In particle–time language: $P_1(t)$ is the x-coordinate of the particle at time t, and $P_2(t)$ is its y-coordinate.

Now comes the key to the proof, certainly not an obvious idea. Since the particle starts at a point in B and ends up at a point in A, there must be a *latest time τ, after which it never returns to B*. We find this time as follows.

Whenever the particle is in B, its x-coordinate is 0, so it is in B for those t for which $P_1(t) = 0$, and for no other times. We can collect these times together as the set $P_1^{-1}(\{0\})$.

Because $\{0\}$ is a closed subset of the x-axis and P_1 is continuous, $P_1^{-1}(\{0\})$ is closed in $[0,1]$. As a bounded subset of $\mathbb{R}$, $P_1^{-1}(\{0\})$ has a least upper bound; call it τ. Since $P_1^{-1}(\{0\})$ is closed, $P_1(\tau) = 0$ and $P(\tau) \in B$, and, for all later times t such that $\tau < t \leq 1$, $P(t) \in A$. Note that $\tau < 1$, since at $t = 1$ the particle is known to be at the point $(1,0)$ in A and not in B.

We now show that $P_2(\tau)$ must be equal to every value in $[0,1]$, giving a contradiction. We do this by observing that, for each $a \in [0,1]$, our assumption that P_2 is continuous implies that

$P_2^{-1}(\{a\})$ is a closed (possibly empty) subset of $[0,1]$.

Thus, if for some $a \in [0,1]$, we show that each neighbourhood of τ intersects $P_2^{-1}(\{a\})$, then

$$\tau \in \operatorname{Cl}(P_2^{-1}(\{a\})) = P_2^{-1}(\{a\}),$$

and so $P_2(\tau) = a$. In fact we show this is true for *all* $a \in [0,1]$.

So, our claim is as follows.

Let $a \in [0,1]$. For each neighbourhood U of τ, there is $t \in U$ for which $P_2(t) = a$.

To verify this claim, we fix $a \in [0,1]$, and let U be a neighbourhood of τ. Then there is $\delta > 0$ such that

$$[\tau, \tau + \delta) \subset U.$$

It suffices to find $s \in [\tau, \tau + \delta)$ for which $P_2(s) = a$.

We noted earlier that

$$0 = P_1(\tau) < P_1(\tau + \delta) \leq 1,$$

and so we can find $N \in \mathbb{N}$ for which

$$0 = P_1(\tau) < \frac{1}{N} < P_1(\tau + \delta) \leq 1.$$

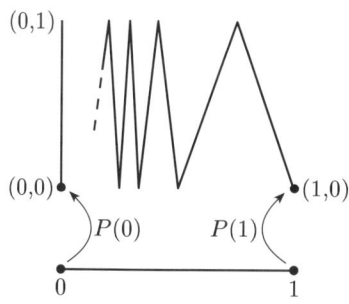

Figure 5.7

We used this language, with a snail instead of a particle, at the start of Section 4.

Note how this example depends crucially on the least upper bound property of $\mathbb{R}$ (Subsection 3.1).

If $P_1(\tau + \delta) = 0$, then τ is not the least upper bound of $P_1^{-1}(\{0\})$.

However P_1 is continuous, and so, by the Intermediate Value Theorem (Theorem 3.3), there is $t_0 \in (\tau, \tau + \delta)$ for which
$$P_1(t_0) = \frac{1}{N},$$
as illustrated in Figure 5.8. Observe that
$$P_2(t_0) = 0.$$

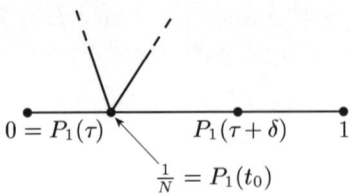

Figure 5.8

Since
$$0 = P_1(\tau) < P_1(t_0) = \frac{1}{N} \quad \text{and} \quad \tfrac{1}{2}\left(\frac{1}{N} + \frac{1}{N+1}\right) < \frac{1}{N},$$
we can again use the Intermediate Value Theorem for P_1 to find $\tau < t_1 < t_0$ for which
$$P_1(t_1) = \tfrac{1}{2}\left(\frac{1}{N} + \frac{1}{N+1}\right),$$
and so $P_2(t_1) = 1$ (see Figure 5.9).

Thus $\tau < t_1 < t_0 < \tau + \delta$, $P_2(t_1) = 1$ and $P_2(t_0) = 0$. But P_2 is continuous, and so we can now use the Intermediate Value Theorem for P_2 to find $t_1 \leq s \leq t_0$ with $P_2(s) = a$. The claim is proved.

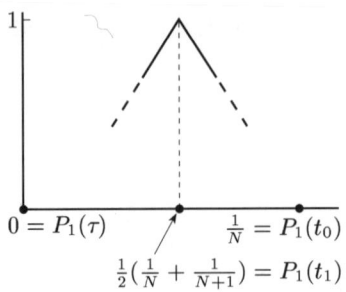

Figure 5.9

This gives our required contradiction, since our earlier observation implies that $\tau \in P_2^{-1}(\{a\})$ for each $a \in [0, 1]$, and so $P_2(\{\tau\}) = [0, 1]$ which is impossible.

Hence C is not path-connected. ∎

We have thus demonstrated that the topologist's cosine is a connected set that is not path-connected.

Solutions to problems

1.1 In each case, we need to find two non-empty disjoint sets, open for the subspace topology, whose union is the whole subset. We make the natural choices.

(a) $\{(-\infty, 0), (1, \infty)\}$.
(b) $\{(-\infty, 0], (1, \infty)\}$.
(c) $\{\{0\}, (1, 2]\}$.

1.2 Since the only subset of $\varnothing$ is $\varnothing$ itself, it follows that there is only one possible topology on $\varnothing$. This topology contains only one set — namely, $\varnothing$ — and so no disconnection exists. So this space is connected.

1.3 The only proper non-empty open sets in $\mathcal{T}_1$ are not disjoint, so there are no disconnections. So $(X, \mathcal{T}_1)$ is connected.
For $(X, \mathcal{T}_2)$, the pair $\{a\}$ and $\{b, c\}$ form a disconnection, so $(X, \mathcal{T}_2)$ is disconnected.

1.4 The only topology that can be defined on a one-element set X is the topology $\{\varnothing, X\}$. Since $\varnothing$ and X are the only possible open subsets, no disconnection of X exists, so $(X, \mathcal{T})$ is connected.

1.5 Since $\varnothing$ and X are the only possible open subsets, no disconnection of X exists, so $(X, \mathcal{T})$ is connected.

1.6 Let q be a rational number. Then $U = (-\infty, q) \cap (\mathbb{R} - \mathbb{Q})$ and $V = (q, \infty) \cap (\mathbb{R} - \mathbb{Q})$ form a $\mathcal{T}_{\mathbb{R}-\mathbb{Q}}$-disconnection of $\mathbb{R} - \mathbb{Q}$. So $(\mathbb{R} - \mathbb{Q}, \mathcal{T}_{\mathbb{R}-\mathbb{Q}})$ is disconnected.

1.7 $\{\{-1\}, (-1, 1]\}$ is a disconnection of $X = [-1, 1]$ and so $(X, \mathcal{T})$ is disconnected. $\{\{1\}, [-1, 1)\}$ is also a disconnection of X, and so is $\{\{1, -1\}, (-1, 1)\}$.

1.8 Assume that a disconnection $\{U, V\}$ exists. Since U and V are non-empty proper open subsets of X, $a \notin U$ and $a \notin V$. Hence the union of U and V does not contain a, and so $U \cup V \neq X$. We conclude that no disconnection exists: $(X, \mathcal{T}_a)$ is connected.

1.9 If $\{U, V\}$ is a disconnection of X we know from Lemma 1.1 that both U and V are closed. Hence $\text{Cl}(U) = U$ and $\text{Cl}(V) = V$, and so $\text{Cl}(U) \cap V = U \cap V = \varnothing$ and $U \cap \text{Cl}(V) = U \cap V = \varnothing$.

1.10 (a)

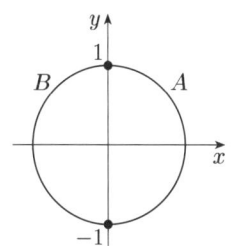

$A \cap B = \{(0, 1), (0, -1)\}$

(b) The sets $U = \{(0, 1)\}$ and $V = \{(0, -1)\}$ are both open for the Euclidean subspace topology on $A \cap B$, and so form a disconnection of $A \cap B$. Thus $A \cap B$ is disconnected.

2.1 There are many possible examples here.
Let $X = \{0, 1\}$ with the discrete topology, so that X is disconnected. Let $Y = \{1\}$ with the indiscrete topology, so that Y is connected. Define $f: X \to Y$ by $f(x) = 1$ for $x \in X$: the map f is continuous by Theorem 4.1 of *Unit A3*.

2.2 We need to find the component of each of the three elements a, b and c of X.

First observe that $\{\{a\}, \{b, c\}\}$ is a disconnection of X, and so X is disconnected. Hence X cannot be the component of any of its elements.

Component of a Since X cannot be the component of a, the only candidates are $\{a\}, \{a, b\}$ and $\{a, c\}$. But $b \notin C_a$, since $\{\{a\}, \{b\}\}$ is a disconnection of $\{a, b\}$ for the subspace topology on $\{a, b\}$. Similarly, $c \notin C_a$. Hence $C_a = \{a\}$.

Component of b Since X and $\{a, b\}$ are disconnected, the only candidates are $\{b\}$ and $\{b, c\}$. The only way to obtain a disconnection of $\{b, c\}$ for the subspace topology would be to find open sets U and V for which $U \cap \{b, c\} = \{b\}$ and $V \cap \{b, c\} = \{c\}$. However,
$$\varnothing \cap \{b, c\} = \varnothing, \quad \{a\} \cap \{b, c\} = \varnothing,$$
$$\{b, c\} \cap \{b, c\} = \{b, c\}, \quad X \cap \{b, c\} = \{b, c\},$$
and so no such sets U and V can exist. Hence $\{b, c\}$ is connected and so $C_b = \{b, c\}$.

Component of c We can deduce from the above analysis that $C_c = \{b, c\}$. We conclude that the components of X are $\{a\}$ and $\{b, c\}$.

2.3 The proof is by contradiction. Let $x \in X$ and assume that C_x consists of more than one element, so there is $y \neq x$ with $y \in C_x$. Then $\{x\}$ and $C_x - \{x\}$ are non-empty, disjoint and open (since every subset is open for the discrete topology). Also, $\{x\} \cup (C_x - \{x\}) = C_x$ and so $\{\{x\}, C_x - \{x\}\}$ is a disconnection of C_x with respect to the subspace topology on C_x. But this is impossible, since C_x is connected. So $C_x = \{x\}$ and hence X is totally disconnected.

2.4 It follows from Theorem 2.4 that $\text{Cl}(C_x)$ is connected. But $\text{Cl}(C_x)$ contains C_x, and C_x is the largest connected set containing x. So $\text{Cl}(C_x) = C_x$, and hence C_x is closed.

2.5 First observe that $\{\{a,b\},\{c,d\}\}$ is a disconnection of X and so X is not connected. Hence X cannot be the component of any of its elements. We know from Lemma 2.7 that the components of X are closed and so they belong to the following collection of sets:
$$\mathcal{T}^c = \{X, \{a,c,d\}, \{a,b,d\}, \{c,d\}, \{a,d\}, \{a,b\},$$
$$\{d\}, \{a\}, \varnothing\}.$$
Now b belongs to only two proper subsets of X in this collection and so $C_b = \{a,b,d\}$ or $C_b = \{a,b\}$. Similarly, $C_c = \{a,c,d\}$ or $C_c = \{c,d\}$.

Since the components of X are also disjoint (by Theorem 2.6), it follows that $C_b = \{a,b\}$ and $C_c = \{c,d\}$. Since the union of these two components is equal to X, it follows from Theorem 2.7 that these are the only components of X.

3.1 Let $B = \{-a : a \in A\}$. Since A is bounded below, B is bounded above, and so B has a least upper bound, M say. Since $b \leq M$ for all $b \in B$, it follows that $-a \leq M$ for all $a \in A$, and so $-M \leq a$ for all $a \in A$. Hence $-M$ is a lower bound for A.

If $m' > -M$, then $-m' < M$, and so there is $b \in B$ for which $-m' < b$. But this means there is $a \in A$ with $-m' < -a$, and this rearranges to give $a < m'$. Hence $-M$ is the greatest lower bound of A.

3.2 Since Theorem 2.2 implies that the continuous image of a connected space is connected, and we have seen in Worked problem 3.1 that $(0,1)$ is connected, it is enough to find a continuous map from $(0,1)$ onto (a,b). One such map is $\phi \colon (0,1) \to (a,b)$, given by
$$\phi(x) = a + (b-a)x.$$
Since this is the restriction to $(0,1)$ of a polynomial (a basic continuous function), it is continuous.

3.3 All these intervals contain (a,b) and are contained in the interval $[a,b]$, the closure of (a,b) in $\mathbb{R}$, and so are connected, by Theorem 2.4.

3.4 The sets U and V are open for the subspace topology on A since they are intersections with A of the Euclidean open sets $\tilde{U}$ and $\tilde{V}$. Neither is empty since $a \in \tilde{U}$ and $b \in \tilde{V}$. The intersection of U and V is
$$U \cap V = (\tilde{U} \cap \tilde{V}) \cap A = \varnothing,$$
and their union is
$$U \cup V = (\tilde{U} \cup \tilde{V}) \cap A = (\mathbb{R} - \{c\}) \cap A = A.$$
Thus $\{U, V\}$ is a disconnection of A.

3.5 The function f is continuous and the space $(\mathbb{R}^2, \mathcal{T}(d^{(2)}))$ is connected, so we can apply the Intermediate Value Theorem. We must find a point in $\mathbb{R}^2$ at which the function takes a value less than 42 and another point at which the function takes a value greater than 42. For example,
$$f(0,0) = 0 \quad \text{and} \quad f(2,0) = 5 \times 2^5 = 160.$$
So, by the Intermediate Value Theorem, there is a point in the plane at which f attains the value 42. (The theorem does not tell us what this point is, just that such a point exists. As we remarked in *Unit A1*, the Intermediate Value Theorem is an existence result.)

3.6 $\{(-\infty, a), (a, \infty)\}$ is a disconnection of $\mathbb{R} - \{a\}$.

3.7 The proof is by contradiction. We know that A and B are both connected. Suppose that there is a homeomorphism $\phi \colon A \to B$.

Let $a \in A$ be the point for which $\phi(a) = 0$ and $b \in A$ be the point for which $\phi(b) = 1$. Then, by the Restriction Rule for homeomorphisms, $A - \{a,b\}$ is homeomorphic to $\phi(A - \{a,b\}) = B - \{0,1\} = (0,1)$. This, however, is a contradiction since $A - \{a,b\}$ is disconnected while $(0,1)$ is connected. Thus A is not homeomorphic to B. (We need to consider both the point a for which $\phi(a) = 0$ and the point b for which $\phi(b) = 1$ to ensure a disconnection here. This is because it is possible that $\phi(1) = 0$ or $\phi(1) = 1$, in which case $A - \{1\} = (0,1)$ which is connected. The second point ensures that we have a point interior to $(0,1)$ with which to disconnect A.)

3.8 Let $(X, \mathcal{T}_X)$ and $(Y, \mathcal{T}_Y)$ be non-empty topological spaces, and suppose that $(X \times Y, \mathcal{T}_X \times \mathcal{T}_Y)$ is connected. We know that the projection functions are continuous, and that $X = p_1(X \times Y)$, $Y = p_2(X \times Y)$. So X and Y are images of a connected space under continuous mappings, and so are connected (by Theorem 2.2).

4.1 Define $f \colon [0,1] \to [0,1] \times [0,1]$ by $f(t) = (t, \tfrac{1}{2}t)$. We note that $(p_1 \circ f)(t) = t$ and $(p_2 \circ f)(t) = \tfrac{1}{2}t$ are continuous functions from $[0,1]$ to $[0,1]$. So, by Theorem 5.7 of *Unit A3*, f is continuous. Hence, since $f(0) = (0,0)$ and $f(1) = (1, \tfrac{1}{2})$, f is a path joining $(0,0)$ and $(1, \tfrac{1}{2})$.

4.2 Define $p_2 \colon [0,1] \to X$ by $p_2(t) = p_1(1-t)$. Then $p_2(0) = p_1(1) = b$ and $p_2(1) = p_1(0) = a$. Moreover, since $p_2 = p_1 \circ f$, where $f \colon [0,1] \to [0,1]$ is the basic continuous function given by $f(t) = 1 - t$, it follows that p_2 is a composite of continuous functions and so is continuous. Hence p_2 is a path in X with initial point b and final point a.

4.3 If $a=b$, we already know that $[a,a]=\{a\}$ is path-connected. We assume without loss of generality that $a<b$.

If $a \neq b$, let $c,d \in [a,b]$.

If $c=d$, we already know how to define a path from c to d, so suppose that $c \neq d$. We assume without loss of generality that $c<d$.

We define $P:[0,1] \to [a,b]$ by
$$P(t) = (1-t)c + td.$$
We know from Worked problem 4.3 that P defines a path from c to d in $\mathbb{R}$. Thus, using Theorem 4.6 of *Unit A3*, we can deduce that P defines a path in $[a,b]$ if $P([0,1]) \subseteq [a,b]$. Now, for $0 \leq t \leq 1$,
$$P(t) = (1-t)c + td = c + (d-c)t \leq c + (d-c) = d.$$
Also, for $0 \leq t \leq 1$, since $c<d$, we have
$$P(t) = c + (d-c)t \geq c.$$
Hence $P([0,1]) \subseteq [c,d] \subseteq [a,b]$, and so P defines a path in $[a,b]$.

Therefore $[a,b]$ is path-connected.

4.4 We must prove that P is a path from $\mathbf{a}$ to $\mathbf{b}$.

To prove continuity, we proceed as follows. For $1 \leq j \leq n$, let $p_j: \mathbb{R}^n \to \mathbb{R}$ be the jth projection function, defined by $p_j(\mathbf{a}) = a_j$. Then $(p_j \circ P)(t) = (1-t)a_j + tb_j$. As in Worked problem 4.3, we deduce that $p_j \circ P: [0,1] \to \mathbb{R}$ is continuous. Since $p_j \circ P$ is continuous for $j = 1, 2, \ldots, n$, it follows from the generalization of Theorem 5.7 of *Unit A3* that P is continuous.

Now $P(0) = \mathbf{a}$ and $P(1) = \mathbf{b}$. Thus P is a path from $\mathbf{a}$ to $\mathbf{b}$. Hence $\mathbb{R}^n$ is path-connected.

4.5 Let a and b be two points in X. We must find a path joining a and b. Since every function $f:[0,1] \to X$ is $(\mathcal{T}(d^{(1)}), \mathcal{T})$-continuous, a path will be defined by any function $p:[0,1] \to X$ such that $p(0) = a$ and $p(1) = b$. For example, we could define p by
$$p(t) = \begin{cases} a & \text{if } t = 0, \\ b & \text{if } 0 < t \leq 1, \end{cases}$$
to obtain a path joining a to b.

Thus $(X, \mathcal{T})$ is path-connected.

4.6 We know, from Worked problem 1.1, that X is not connected. It follows from Theorem 4.2 that X cannot be path-connected.

4.7 For any $x \in [0,1]$, and any $s,t \in [0,1]$,
$$\begin{aligned}|P(t)(x) - P(s)(x)| &= |(1-t)f(x) + tg(x) \\ &\quad - (1-s)f(x) - sg(x)| \\ &= |(t-s)(g(x) - f(x))| \\ &= |t-s||g(x) - f(x)| \\ &\leq |t-s| d_{\max}(f,g). \end{aligned}$$
Taking the maximum over $x \in [0,1]$, we have
$$d_{\max}(P(t), P(s)) \leq |t-s| d_{\max}(f,g).$$
We prove that P is continuous at $t \in [0,1]$.

Given any $\varepsilon > 0$, choose $\delta > 0$ such that
$$\delta < \varepsilon / (d_{\max}(f,g) + 1).$$
Then, if $s \in [0,1]$ is such that $|t-s| < \delta$,
$$d_{\max}(P(t), P(s)) < \delta\, d_{\max}(f,g) < \varepsilon.$$
Thus, P is continuous at every $t \in [0,1]$.

4.8 By Theorem 6.2 of *Unit A3*, if $p:[0,1] \to X$ is a path when X carries the topology $\mathcal{T}_1$, it remains continuous when X carries the coarser topology $\mathcal{T}_2$, and so is a path in $(X, \mathcal{T}_2)$.

Index

clopen set, 6
closed path, 24
component, 13
connected, 18
    -path, 26
    Euclidean space, 16
    interval, 18
    product space, 21
    set, 5
    space, 5

disconnected
    set, 5
    space, 5
disconnection, 5

final point, 24

greatest lower bound, 16

homeomorphic invariance
    connectedness, 12
    path-connectedness, 31

infimum, 16
initial point, 24
Intermediate Value Theorem, 19
interval, 18
    classes, 20

least upper bound, 16

path
    closed, 24
    final point, 24
    initial point, 24
path-connected
    product space, 30
    set, 26
    space, 26
proper subset, 6

supremum, 16

topologist's cosine, 32
totally disconnected, 13
trail, 24